SpringerBriefs in Mathematics

SpringerBriefs in Mathematics showcases expositions in all areas of mathematics and applied mathematics. Manuscripts presenting new results or a single new result in a classical field, new field, or an emerging topic, applications, or bridges between new results and already published works, are encouraged. The series is intended for mathematicians and applied mathematicians.

More information about this series at http://www.springer.com/series/10030

Saber Jafarpour • Andrew D. Lewis

Time-Varying Vector Fields and Their Flows

Springer

Saber Jafarpour
Department of Mathematics and Statistics
Queen's University
Kingston, ON, Canada

Andrew D. Lewis
Department of Mathematics and Statistics
Queen's University
Kingston, ON, Canada

ISSN 2191-8198 ISSN 2191-8201 (electronic)
ISBN 978-3-319-10138-5 ISBN 978-3-319-10139-2 (eBook)
DOI 10.1007/978-3-319-10139-2
Springer Cham Heidelberg New York Dordrecht London

Library of Congress Control Number: 2014952196

Mathematics Subject Classification: 32C05, 34A12, 46E10

Printed on acid-free paper

Springer is part of Springer Science+Business Media (www.springer.com)

Preface

In this monograph, time-varying vector fields on manifolds are considered with measurable time dependence and with varying degrees of regularity in state; e.g., Lipschitz, finitely differentiable, smooth, holomorphic, and real analytic. Classes of such vector fields are described for which the regularity of the flow's dependence on initial condition matches the regularity of the vector field. While results of this sort are known for continuous dependence, some results exist for differentiable dependence, and smooth dependence seems to be a part of the folklore, such results have certainly not been established in the important real analytic case. We treat *all* regularity classes, including the real analytic case. Moreover, it is shown that the way in which one characterises these classes of vector fields corresponds exactly to the vector fields, thought of as being vector field valued functions of time, having measurability and integrability properties associated with appropriate topologies. For this reason, a substantial part of the development is concerned with descriptions of these appropriate topologies. To this end, geometric descriptions are provided of locally convex topologies for Lipschitz, finitely differentiable, smooth, holomorphic, and real analytic sections of vector bundles. In all but the real analytic case, these topologies are classically known. The description given for the real analytic topology is new. It is this description that allows, for the first time, a characterisation of those time-varying real analytic vector fields whose flows depend real analytically on initial condition.

This research was funded in part by a grant from the Natural Sciences and Engineering Research Council of Canada. The second author was a Visiting Professor in the Department of Mathematics at University of Hawaii, Manoa, when the monograph was written, and would like to acknowledge the hospitality of the department, particularly that of Monique Chyba and George Wilkens. The second author would also like to thank his departmental colleague Mike Roth for numerous useful conversations over the years. While conversations with Mike did not lead directly to results in this monograph, Mike's willingness to chat about complex geometry and to answer ill-informed questions was always appreciated and ultimately very helpful.

Kingston, ON, Canada Saber Jafarpour
Kingston, ON, Canada Andrew D. Lewis

Contents

Chapter 1
Introduction

1.1 Motivation

In this monograph we consider vector fields with measurable time dependence. Such vector fields are not well studied in the differential equation and dynamical system literature, but are important in control theory. There are at least two reasons for this:

1. In the theory of optimal control, it can often happen that optimal trajectories correspond to controls that switch infinitely often in a finite duration of time. This seems to have first been observed by Fuller [16], and has since been studied by many authors. A detailed discussion of these facets of optimal control theory can be found in [41].
2. In the study of controllability, it is often necessary to use complex control variations where there are an increasing number of switches in a finite time. This was noticed by Kawski [28] and examined in detail by Agrachev and Gamkrelidze [3].

In any event, vector fields with measurable time dependence do not have as extensive a basic theory as, say, time-independent vector fields. One place where this especially holds true is where regularity of flows with respect to initial conditions is concerned. For time-independent vector fields, the basic regularity theorems are well established for all the standard degrees of regularity [9], including real analyticity, e.g., [38, Proposition C.3.12]. In the time-varying case, the standard Carathéodory existence and uniqueness theorem for ordinary differential equations depending measurably on time is a part of classical text treatments, e.g., [9, Theorem 2.2.1]. Continuity with respect to initial condition in this setting is not consistently mentioned, e.g., it is not proved by Coddington and Levinson [9]. This continuity is proved, for example, by Sontag [38, Theorem 55]. However, conditions for differentiability of flows become extremely hard to come by in the case of measurable time dependence. The situation here, however, is dealt with comprehensively in [37]. The differentiable hypotheses are easily extended to any finite degree of differentiability. For smooth dependence on initial conditions with measurable time dependence, we are not aware of the desired result being stated *and proved*

S. Jafarpour, A.D. Lewis, *Time-Varying Vector Fields and Their Flows*,
SpringerBriefs in Mathematics, DOI 10.1007/978-3-319-10139-2_1

anywhere in the existing literature. We prove this result as our Theorem 6.6, and we note that the hypotheses guaranteeing smooth dependence on initial condition are what one might guess after a moment's reflection, and understanding the conditions for differentiable dependence. However, the situation is quite different for vector fields depending on state in a real analytic manner. In this case, it is by no means clear a priori what are the correct hypotheses for such a vector field to have a flow depending on initial condition in a real analytic manner. Indeed, the joint conditions on time and state required to ensure such real analytic dependence are simply not known, and are being given here for the first time.

Quite apart from the sort of pragmatic issue of determining the appropriate hypotheses for real analytic dependence on initial conditions, what we reveal in this work is that the matter of joint conditions on state and time that give desired regularity with respect to initial conditions are intimately connected with topologies for spaces of vector fields. Let us indicate the nature of this connection. Let M be a manifold and suppose that we have a time-varying vector field $(t, x) \mapsto X(t, x) \in \mathsf{TM}$. This then defines a curve $t \mapsto X_t$ in the space of vector fields, with $X_t(x) = X(t, x)$. If the space of vector fields has a topology, then one can consider properties like measurability, integrability, and boundedness of a curve such as this. We show that the characterisations of these attributes of curves of vector fields are precisely related to the hypotheses required to prove regular dependence on initial conditions. That is, we reveal the connection between regular dependence on initial conditions and descriptions of topologies for spaces of vector fields. We show that this is the case for *all* degrees of regularity, and so, in particular, we see that the standard conditions for existence, uniqueness, and regularity are, in fact, measurability and integrability conditions in appropriate spaces of vector fields.

For this reason, we spend a great deal of time describing topologies for spaces of sections of vector bundles. We do this for spaces of Lipschitz, finitely differentiable, smooth, holomorphic, and real analytic sections of vector bundles. These topologies are classical in the finitely differentiable, smooth, and holomorphic cases. The extension to the Lipschitz case is fairly easily carried out, but we are unaware of this being done in the literature, so we provide this extension here. But the main contribution is a description of the topology of the space of real analytic sections. We do this by providing explicit seminorms for this topology. The seminorms we use in the real analytic case are defined in such a way that the qualitative relationships between the real analytic and the other regularity classes are clear. For this reason, as part of our presentation we provide a thorough treatment of the Lipschitz, finitely differentiable, smooth, and holomorphic cases alongside the novel real analytic presentation. This serves to illustrate the unified framework that we have developed for analysing time-varying vector fields with measurable time dependence and varying degrees of regularity in state. Indeed, one of the satisfying aspects of the way the theory is developed is that many things "look the same" for all degrees of regularity; only the seminorms change. That being said, the main contribution is the explication of the real analytic theory, and we shall see that there are a few places where this case is substantially more difficult than the other regularity classes we consider.

The ideas we present in this work have their origins in the work of Agrachev and Gamkrelidze [2] on "chronological calculus". Our work here is distinguished, however, in two important ways. First of all, the work in [2] is presented in Euclidean space, while we work in a global framework of vector fields on manifolds. In the presentation of chronological calculus in the recent book of Agrachev and Sachkov [4], the formulation is also given on manifolds, but the analysis methods, particularly the seminorms used, use an embedding of the manifold in Euclidean space by the Whitney Embedding Theorem [40]. In contrast, we use seminorms formulated intrinsically in terms of fibre norms on jet bundles. This use of geometric seminorms allows for an elegant and unified treatment of all regularity classes, with many of the fundamental theorems having hypotheses closely resembling one another, and having proofs that rely on properties of topologies that are shared by all regularity classes. A second significant difference in our approach and that of [2] is the manner in which the real analytic case is handled. In this latter work, the real analytic analysis is restricted to real analytic vector fields defined on real Euclidean space admitting a bounded holomorphic extension to a neighbourhood of fixed width in complex Euclidean space. This is a rather severe restriction, and one that we eliminate by defining an appropriate topology for the space of real analytic vector fields. We comment that our description of this topology using geometric seminorms is only made possible by the recent study of real analytic analysis by, e.g., [13, 14, 39]. That is to say, the complete results we give here are only possible due to work that was not available at the time of [2]. It is fair to say that our work here is a completion—a nontrivial completion—of the project initiated in this earlier work.

1.2 An Outline of the Monograph

Let us discuss briefly the contents of the monograph.

One of the essential elements of this work is a characterisation of seminorms for the various topologies we use. Our definitions of these seminorms unify the presentation of the various degrees of regularity we consider—finitely differentiable, Lipschitz, smooth, holomorphic, and real analytic—making it so that, after the seminorms are in place, these various cases can be treated in very similar ways in many cases. The key to the construction of the seminorms that we use is the use of connections to decompose jet bundles into direct sums. In Chap. 2 we present these constructions. As we see in Chap. 5, in the real analytic case, some careful estimates must be performed to ensure that the geometric seminorms we use do, indeed, characterise the real analytic topology. These estimates are carried out in Chap. 2.

In Chaps. 3, 4, and 5 we describe topologies for spaces of finitely differentiable, Lipschitz, smooth, holomorphic, and real analytic vector fields. While these topologies are more or less classical in the smooth, finitely differentiable, and holomorphic cases, in the real analytic case the description we give is less well known, and indeed many of our results here are new, or provide new and useful ways of understanding existing results. We also fully develop various "weak" formulations of properties such as continuity, boundedness, measurability, and integrability for spaces of finitely differentiable, Lipschitz, smooth, and real analytic vector fields. In Chaps. 3, 4, and 5 the weak formulations we develop are concerned with evaluations of vector fields on functions by Lie differentiation, which we call the "weak-$\mathscr{L}$" topology. In the existing literature, the weak-$\mathscr{L}$ topology is often used with some sort of implicit understanding that it is equivalent to its "strong" counterpart. Here we validate this implicit understanding with explicit proofs for all regularity classes.

In Chap. 6 we present our main results concerning time-varying vector fields. In the smooth case, the ideas we present are probably contained in the work of Agrachev and Gamkrelidze [2] (see also [4]), but our presentation of the real analytic case is novel. For this reason, we present a rather complete treatment of the smooth case (with the finitely differentiable and Lipschitz cases following along similar lines) so as to provide a context for the more complicated real analytic case. We should point out that, even in the smooth case, we use properties of the topology that are not normally called upon, and we see that it is these deeper properties that really tie together the various regularity hypotheses we use. Indeed, what our presentation reveals is the connection between the standard pointwise—in time and state—conditions placed on time-varying vector fields and topological characterisations. This is, we believe, a fulfilling way of understanding the meaning of the usual pointwise conditions.

1.3 Notation, Conventions and Background

In this section we overview what is needed to read the monograph. We do use a lot of specialised material in essential ways, and we certainly do not review this comprehensively. Instead, we simply provide a few facts, the notation we shall use, and recommended sources. Throughout the work we have tried to include precise references to material needed so that a reader possessing enthusiasm and lacking background can begin to chase down all of the ideas upon which we rely.

We shall use the slightly unconventional, but perfectly rational, notation of writing $A \subseteq B$ to denote set inclusion, and when we write $A \subset B$ we mean that $A \subseteq B$ and $A \neq B$. By id_A we denote the identity map on a set A. For a product $\prod_{i \in I} X_i$ of sets, $\mathrm{pr}_j \colon \prod_{i \in I} X_i \to X_j$ is the projection onto the jth component. For a subset $A \subseteq X$, we denote by χ_A the characteristic function of A, i.e.,

$$\chi_A(x) = \begin{cases} 1, & x \in A, \\ 0, & x \notin A. \end{cases}$$

By $\mathrm{card}(A)$ we denote the cardinality of a set A. By $\mathfrak{S}_k$ we denote the symmetric group on k symbols. By $\mathbb{Z}$ we denote the set of integers, with $\mathbb{Z}_{\geq 0}$ denoting the set of nonnegative integers and $\mathbb{Z}_{>0}$ denoting the set of positive integers. We denote by $\mathbb{R}$ and $\mathbb{C}$ the sets of real and complex numbers. By $\mathbb{R}_{\geq 0}$ we denote the set of nonnegative real numbers and by $\mathbb{R}_{>0}$ the set of positive real numbers. By $\overline{\mathbb{R}}_{\geq 0} = \mathbb{R}_{\geq 0} \cup \{\infty\}$ we denote the extended nonnegative real numbers. By δ_{jk}, $j, k \in \{1, \dots, n\}$, we denote the Kronecker delta.

We shall use constructions from algebra and multilinear algebra, referring to [26], [6, Chap. III], and [7, Sect. IV.5]. If F is a field (for us, typically $\mathsf{F} \in \{\mathbb{R}, \mathbb{C}\}$), if V is an F-vector space, and if $A \subseteq \mathsf{V}$, by $\operatorname{span}_{\mathsf{F}}(A)$ we denote the subspace generated by A. If F is a field and if U and V are F-vector spaces, by $\operatorname{Hom}_{\mathsf{F}}(\mathsf{U}; \mathsf{V})$ we denote the set of linear maps from U to V. We denote $\operatorname{End}_{\mathsf{F}}(\mathsf{V}) = \operatorname{Hom}_{\mathsf{F}}(\mathsf{V}; \mathsf{V})$ and $\mathsf{V}^* = \operatorname{Hom}_{\mathsf{F}}(\mathsf{V}; \mathsf{F})$. If $\alpha \in \mathsf{V}^*$ and $v \in \mathsf{V}$, we may sometimes denote by $\langle \alpha; v \rangle \in \mathsf{F}$ the natural pairing. The k-fold tensor product of V with itself is denoted by $\mathrm{T}^k(\mathsf{V})$. Thus, if V is finite-dimensional, we identify $\mathrm{T}^k(\mathsf{V}^*)$ with the k-multilinear F-valued functions on V^k by

$$(\alpha^1 \otimes \cdots \otimes \alpha^k)(v_1, \dots, v_k) = \alpha^1(v_1) \cdots \alpha^k(v_k).$$

By $\mathrm{S}^k(\mathsf{V}^*)$ we denote the symmetric tensor algebra of degree k, which we identify with the symmetric k-multilinear F-valued functions on V^k, or polynomial functions of homogeneous degree k on V.

If $\mathbb{G}$ is an inner product on a $\mathbb{R}$-vector space V, we denote by $\mathbb{G}^\flat \in \operatorname{Hom}_{\mathbb{R}}(\mathsf{V}; \mathsf{V}^*)$ the associated mapping and by $\mathbb{G}^\sharp \in \operatorname{Hom}_{\mathbb{R}}(\mathsf{V}^*; \mathsf{V})$ the inverse of $\mathbb{G}^\flat$.

Elements of $\mathbb{F}^n$, $\mathbb{F} \in \{\mathbb{R}, \mathbb{C}\}$, are typically denoted with a bold font, e.g., "$\boldsymbol{x}$". The standard basis for $\mathbb{F}^n$ is denoted by $(\boldsymbol{e}_1, \dots, \boldsymbol{e}_n)$. By $\boldsymbol{I}_n$ we denote the $n \times n$ identity matrix. We denote by $\mathrm{L}(\mathbb{R}^n; \mathbb{R}^m)$ the set of linear maps from $\mathbb{R}^n$ to $\mathbb{R}^m$ (this is the same as $\operatorname{Hom}_{\mathbb{R}}(\mathbb{R}^n; \mathbb{R}^m)$, of course, but the more compact notation is sometimes helpful). The invertible linear maps on $\mathbb{R}^n$ we denote by $\mathsf{GL}(n; \mathbb{R})$. By $\mathrm{L}(\mathbb{R}^{n_1}, \dots, \mathbb{R}^{n_k}; \mathbb{R}^m)$ we denote the set of multilinear mappings from $\prod_{j=1}^k \mathbb{R}^{n_j}$ to $\mathbb{R}^m$. We abbreviate by $\mathrm{L}^k(\mathbb{R}^n; \mathbb{R}^m)$ the k-multilinear maps from $(\mathbb{R}^n)^k$ to $\mathbb{R}^m$. We denote by $\mathrm{L}^k_{\mathrm{sym}}(\mathbb{R}^n; \mathbb{R}^m)$ the set of symmetric k-multilinear maps from $(\mathbb{R}^n)^k$ to $\mathbb{R}^m$. With our notation above, $\mathrm{L}^k_{\mathrm{sym}}(\mathbb{R}^n; \mathbb{R}^m) \simeq \mathrm{S}^k((\mathbb{R}^n)^*) \otimes \mathbb{R}^m$, but, again, we prefer the slightly more compact notation in this special case.

For a topological space $\mathcal{X}$ and $A \subseteq \mathcal{X}$, $\operatorname{int}(A)$ denotes the interior of A and $\operatorname{cl}(A)$ denotes the closure of A. Neighbourhoods will always be open sets.

By $\mathsf{B}(r, \boldsymbol{x}) \subseteq \mathbb{R}^n$ we denote the open ball of radius r and centre $\boldsymbol{x}$. If $r \in \mathbb{R}_{>0}$ and if $x \in \mathbb{F}$, $\mathbb{F} \in \{\mathbb{R}, \mathbb{C}\}$, we denote by

$$\mathsf{D}(r, x) = \{x' \in \mathbb{F} \mid |x' - x| < r\}$$

the disk of radius r centred at x. If $\boldsymbol{r} \in \mathbb{R}^n_{>0}$ and if $\boldsymbol{x} \in \mathbb{F}^n$, we denote by

$$\mathsf{D}(\boldsymbol{r}, \boldsymbol{x}) = \mathsf{D}(r_1, x_1) \times \cdots \times \mathsf{D}(r_n, x_n)$$

the polydisk with radius $\boldsymbol{r}$ centred at $\boldsymbol{x}$. In like manner, $\overline{\mathsf{D}}(\boldsymbol{r}, \boldsymbol{x})$ denotes the closed polydisk.

If $\mathcal{U} \subseteq \mathbb{R}^n$ is open and if $\boldsymbol{\Phi}\colon \mathcal{U} \to \mathbb{R}^m$ is differentiable at $\boldsymbol{x} \in \mathcal{U}$, we denote its derivative by $\boldsymbol{D\Phi}(\boldsymbol{x})$. Higher-order derivatives, when they exist, are denoted by

$\boldsymbol{D}^r\boldsymbol{\Phi}(\boldsymbol{x})$, r being the order of differentiation. We will also use the following partial derivative notation. Let $\mathcal{U}_j \subseteq \mathbb{R}^{n_j}$ be open, $j \in \{1,\dots,k\}$, and let $\boldsymbol{\Phi}\colon \mathcal{U}_1 \times \cdots \times \mathcal{U}_k \to \mathbb{R}^m$ be continuously differentiable. The derivative of the map

$$\boldsymbol{x}_j \mapsto \boldsymbol{\Phi}(\boldsymbol{x}_{1,0},\dots,\boldsymbol{x}_j,\dots,\boldsymbol{x}_{k,0})$$

at $\boldsymbol{x}_{j,0}$ is denoted by $\boldsymbol{D}_j\boldsymbol{\Phi}(\boldsymbol{x}_{1,0},\dots,\boldsymbol{x}_{k,0})$. Higher-order partial derivatives, when they exist, are denoted by $\boldsymbol{D}_j^r\boldsymbol{\Phi}(\boldsymbol{x}_{1,0},\dots,\boldsymbol{x}_{k,0})$, r being the order of differentiation. We recall that if $\boldsymbol{\Phi}\colon \mathcal{U} \to \mathbb{R}^m$ is of class C^k, $k \in \mathbb{Z}_{>0}$, then $\boldsymbol{D}^k\boldsymbol{\Phi}(\boldsymbol{x})$ is symmetric. We shall sometimes find it convenient to use multi-index notation for derivatives. A ***multi-index*** with length n is an element of $\mathbb{Z}^n_{\geq 0}$, i.e., an n-tuple $I = (i_1,\dots,i_n)$ of nonnegative integers. If $\boldsymbol{\Phi}\colon \mathcal{U} \to \mathbb{R}^m$ is a smooth function, then we denote

$$\boldsymbol{D}^I\boldsymbol{\Phi}(\boldsymbol{x}) = \boldsymbol{D}_1^{i_1}\cdots\boldsymbol{D}_n^{i_n}\boldsymbol{\Phi}(\boldsymbol{x}).$$

We will use the symbol $|I| = i_1+\cdots+i_n$ to denote the order of the derivative. Another piece of multi-index notation we shall use is

$$\boldsymbol{a}^I = a_1^{i_1}\cdots a_n^{i_n},$$

for $\boldsymbol{a} \in \mathbb{R}^n$ and $I \in \mathbb{Z}^n_{\geq 0}$. Also, we denote $I! = i_1!\cdots i_n!$.

Our differential geometric conventions mostly follow [1]. Whenever we write "manifold", we mean "second-countable Hausdorff manifold". This implies, in particular, that manifolds are assumed to be metrisable [1, Corollary 5.5.13]. If we use the letter "n" without mentioning what it is, it is the dimension of the connected component of the manifold M with which we are working at that time. The tangent bundle of a manifold M is denoted by $\pi_{\mathsf{TM}}\colon \mathsf{TM} \to \mathsf{M}$ and the cotangent bundle by $\pi_{\mathsf{T^*M}}\colon \mathsf{T^*M} \to \mathsf{M}$. The derivative of a differentiable map $\Phi\colon \mathsf{M} \to \mathsf{N}$ is denoted by $T\Phi\colon \mathsf{TM} \to \mathsf{TN}$, with $T_x\Phi = T\Phi|\mathsf{T}_x\mathsf{M}$. If $I \subseteq \mathbb{R}$ is an interval and if $\xi\colon I \to \mathsf{M}$ is a curve that is differentiable at $t \in I$, we denote the tangent vector field to the curve at t by $\xi'(t) = T_t\xi(1)$. We use the symbols Φ^* and Φ_* for pull-back and push-forward. Precisely, if g is a function on N, $\Phi^*g = g\circ\Phi$, and if $\Phi\colon \mathsf{M} \to \mathsf{N}$ is a diffeomorphism, if f is a function on M, if X is a vector field on M, and if Y is a vector field on N, we have $\Phi_*f = f\circ\Phi^{-1}$, $\Phi_*X = T\Phi\circ X\circ\Phi^{-1}$, and $\Phi^*Y = T\Phi^{-1}\circ Y\circ\Phi$. The flow of a vector field X is denoted by Φ_t^X, so $t \mapsto \Phi_t^X(x)$ is the integral curve of X passing through x at $t = 0$. We shall also use time-varying vector fields, but will develop the notation for the flows of these in the text.

If $\pi\colon \mathsf{E} \to \mathsf{M}$ is a vector bundle, we denote the fibre over $x \in \mathsf{M}$ by E_x and we sometimes denote by 0_x the zero vector in E_x. If $\mathsf{S} \subseteq \mathsf{M}$ is a submanifold, we denote by $\mathsf{E}|\mathsf{S}$ the restriction of E to S which we regard as a vector bundle over S. The ***vertical subbundle*** of E is the subbundle of TE defined by $\mathsf{VE} = \ker(T\pi)$. If $\mathbb{G}$ is a fibre metric on E, i.e., a smooth assignment of an inner product to each of the fibres of E, then $\|\cdot\|_{\mathbb{G}}$ denotes the norm associated with the inner product on fibres. If $\pi\colon \mathsf{E} \to \mathsf{M}$ is a vector bundle and if $\Phi\colon \mathsf{N} \to \mathsf{M}$ is a smooth map, then $\Phi^*\pi\colon \Phi^*\mathsf{E} \to \mathsf{N}$ denotes the pull-back of E to N [30, Sect. III.9.5]. The dual of a vector bundle $\pi\colon \mathsf{E} \to \mathsf{M}$ is denoted by $\pi^*\colon \mathsf{E}^* \to \mathsf{M}$.

Generally we will try hard to avoid coordinate computations. However, they are sometimes unavoidable and we will use the Einstein summation convention when it is convenient to do so, but we will not do so slavishly.

We will work in both the smooth and real analytic categories, with occasional forays into the holomorphic category. We will also work with finitely differentiable objects, i.e., objects of class C^r for $r \in \mathbb{Z}_{\geq 0}$. (We will also work with Lipschitz objects, but will develop the notation for these in the text.) A good reference for basic real analytic analysis is [31], but we will need ideas going beyond those from this text, or any other text. Relatively recent work of e.g., [13, 14, 39], has shed a great deal of light on real analytic analysis, and we shall take advantage of this work. An analytic manifold or mapping will be said to be of ***class*** $\boldsymbol{C^\omega}$. Let $r \in \mathbb{Z}_{\geq 0} \cup \{\infty, \omega\}$. The set of mappings of class C^r between manifolds M and N is denoted by $\mathrm{C}^r(\mathsf{M};\mathsf{N})$. We abbreviate $\mathrm{C}^r(\mathsf{M}) = \mathrm{C}^r(\mathsf{M};\mathbb{R})$. The set of sections of a vector bundle $\pi\colon \mathsf{E} \to \mathsf{M}$ of class C^r is denoted by $\Gamma^r(\mathsf{E})$. Thus, in particular, $\Gamma^r(\mathsf{TM})$ denotes the set of vector fields of class C^r. We shall think of $\Gamma^r(\mathsf{E})$ as a $\mathbb{R}$-vector space with the natural pointwise addition and scalar multiplication operations. If $f \in \mathrm{C}^r(\mathsf{M})$, $\mathrm{d}f \in \Gamma^r(\mathsf{T}^*\mathsf{M})$ denotes the differential of f. If $X \in \Gamma^r(\mathsf{TM})$ and $f \in \mathrm{C}^r(\mathsf{M})$, we denote the Lie derivative of f with respect to X by $\mathscr{L}_X f$.

We also work with holomorphic, i.e., complex analytic, manifolds and associated geometric constructions; real analytic geometry, at some level, seems to unavoidably rely on holomorphic geometry. A nice overview of holomorphic geometry, and some of its connections to real analytic geometry, is given in the book [8]. There are many specialised texts on the subject of holomorphic geometry, including [11, 15, 21, 24] and the three volumes of Gunning [18, 19, 20]. For our purposes, we shall just say the following things. By TM we denote the holomorphic tangent bundle of M. This is the object which, in complex differential geometry, is commonly denoted by $\mathsf{T}^{1,0}\mathsf{M}$. For holomorphic manifolds M and N, we denote by $\mathrm{C}^{\mathrm{hol}}(\mathsf{M};\mathsf{N})$ the set of holomorphic mappings from M to N, by $\mathrm{C}^{\mathrm{hol}}(\mathsf{M})$ the set of holomorphic functions on M (note that these functions are $\mathbb{C}$-valued, not $\mathbb{R}$-valued, of course), and by $\Gamma^{\mathrm{hol}}(\mathsf{E})$ the space of holomorphic sections of an holomorphic vector bundle $\pi\colon \mathsf{E} \to \mathsf{M}$. We shall use both the natural $\mathbb{C}$- and, by restriction, $\mathbb{R}$-vector space structures for $\Gamma^{\mathrm{hol}}(\mathsf{E})$.

We will make use of the notion of a "Stein manifold". For practical purposes, these can be taken to be holomorphic manifolds admitting a proper holomorphic embedding in complex Euclidean space.[1] Stein manifolds are characterised by having lots of holomorphic functions, distinguishing them from general holomorphic manifolds, e.g., compact holomorphic manifolds whose only holomorphic functions are those that are locally constant. There is a close connection between Stein mani-

[1] The equivalence of this to other characterisations of Stein manifolds is due to Remmert [33]. A reader unfamiliar with holomorphic manifolds should note that, unlike in the smooth or real analytic cases, it is *not* generally true that an holomorphic manifold can be embedded in complex Euclidean space, even after the usual elimination of topological pathologies such as non-paracompactness. For example, compact holomorphic manifolds can never be holomorphically embedded in complex Euclidean space.

folds and real analytic manifolds, and this explains our interest in Stein manifolds. We shall point out these connections as they arise in the text.

We shall make reference to germs of functions and sections, for which we use the following notation. Let $r \in \mathbb{Z}_{\geq 0} \cup \{\infty, \omega, \mathrm{hol}\}$ and let M be a smooth, real analytic, or holomorphic manifold, such as is demanded by r. By $\mathscr{C}^r_{\mathsf{M}}$ we denote the sheaf of functions of class C^r and by $\mathscr{C}^r_{x,\mathsf{M}}$ the set of germs of this sheaf at $x \in \mathsf{M}$. If $\pi\colon \mathsf{E} \to \mathsf{M}$ is a C^r-vector bundle, then $\mathscr{G}^r_{\mathsf{E}}$ denotes the sheaf of C^r-sections of E with $\mathscr{G}^r_{x,\mathsf{E}}$ the set of germs at x. The germ of a function (resp. section) at x will be denoted by $[f]_x$ (resp. $[\xi]_x$).

We will make use of jet bundles, and a standard reference is [35]. Appropriate sections of [30] (especially Sect. 12) are also useful. If $\pi\colon \mathsf{E} \to \mathsf{M}$ is a vector bundle and if $k \in \mathbb{Z}_{\geq 0}$, we denote by $\mathsf{J}^k\mathsf{E}$ the bundle of k-jets of E. For a section ξ of E, we denote by $j_k\xi$ the corresponding section of $\mathsf{J}^k\mathsf{E}$. The projection from $\mathsf{J}^k\mathsf{E}$ to $\mathsf{J}^l\mathsf{E}$, $l \leq k$, is denoted by π^k_l. If M and N are manifolds, we denote by $\mathsf{J}^k(\mathsf{M};\mathsf{N})$ the bundle of k jets of mappings from M to N. If $\Phi \in \mathrm{C}^\infty(\mathsf{M};\mathsf{N})$, $j_k\Phi$ denotes its k-jet, which is a mapping from M to $\mathsf{J}^k(\mathsf{M};\mathsf{N})$. In the proof of Theorem 6.6 we will briefly make use of jets of sections of fibred manifolds. We shall introduce there the notation we require, and the reader can refer to [35] to fill in the details.

We shall make use of connections and refer to [30, Sects. 11, 17] for a comprehensive treatment of these, or to [29] for another comprehensive treatment and an alternative point of view.

We shall make frequent and essential use of nontrivial facts about locally convex topological vector spaces, and refer to [10, 17, 25, 27, 34, 36] for details. We shall also access the contemporary research literature on locally convex spaces and will indicate this as we go along. In this monograph we shall suppose that locally convex spaces are Hausdorff. We shall denote by $\mathrm{L}(\mathsf{U};\mathsf{V})$ the set of continuous linear maps from a locally convex space U to a locally convex space V. In particular, U' is the topological dual of U, meaning the continuous linear scalar-valued functions. We will break with the usual language one sees in the theory of locally convex spaces and call what are commonly called "inductive" and "projective" limits, instead "direct" and "inverse" limits, in keeping with the rest of category theory. We shall make occasional reference to the notion of a "nuclear locally convex space". There are several ways of characterising nuclear spaces. Here is one. A continuous linear mapping $L\colon \mathsf{E} \to \mathsf{F}$ between Banach spaces is ***nuclear*** if there exist sequences $(v_j)_{j\in\mathbb{Z}_{>0}}$ in F and $(\alpha_j)_{j\in\mathbb{Z}_{>0}}$ in E' such that $\sum_{j\in\mathbb{Z}_{>0}} \|\alpha_j\|\|v_j\| < \infty$ and such that

$$L(u) = \sum_{j=1}^{\infty} \alpha_j(u)v_j,$$

the sum converging in the topology of V. Now suppose that V is a locally convex space and p is a continuous seminorm on V. We denote by $\overline{\mathsf{V}}_p$ the completion of

$$\mathsf{V}/\{v \in \mathsf{V} \mid p(v) = 0\};$$

thus $\overline{\mathsf{V}}_p$ is a Banach space. The space V is ***nuclear*** if, for any continuous seminorm p, there exists a continuous seminorm q satisfying $q \leq p$ such that the mapping

$$\begin{aligned} i_{p,q} \colon \overline{\mathsf{V}}_p &\to \overline{\mathsf{V}}_q \\ v + \{v' \in \mathsf{V} \mid p(v) = 0\} &\mapsto v + \{v' \in \mathsf{V} \mid q(v) = 0\} \end{aligned}$$

is nuclear. It is to be understood that this definition is essentially meaningless at a first encounter, so we refer to [23, 32] and relevant sections of [27] to begin understanding the notion of a nuclear space. The only attribute of nuclear spaces of interest to us here is that their relatively compact subsets are exactly the von Neumann bounded subsets [32, Proposition 4.47].

By λ we denote the Lebesgue measure on $\mathbb{R}$. We will talk about measurability of maps taking values in topological spaces. If $(\mathcal{T}, \mathscr{M})$ is a measurable space and if $\mathcal{X}$ is a topological space, a mapping $\Psi \colon \mathcal{T} \to \mathcal{X}$ is ***Borel measurable*** if $\Psi^{-1}(\mathcal{O}) \in \mathscr{M}$ for every open set $\mathcal{O} \subseteq \mathcal{X}$. This is equivalent to requiring that $\Psi^{-1}(\mathcal{B}) \in \mathscr{M}$ for every Borel subset $\mathcal{B} \subseteq \mathcal{X}$.

One not completely standard topic we shall need to understand is integration of functions with values in locally convex spaces. There are multiple theories here,[2] so let us outline what we mean, following [5]. We let $(\mathcal{T}, \mathscr{M}, \mu)$ be a finite measure space, let V be a locally convex topological vector space, and let $\Psi \colon \mathcal{T} \to \mathsf{V}$. Measurability of Ψ is Borel measurability mentioned above, and we note that there are other forms of measurability that arise for locally convex spaces (the comment made in footnote 2 applies to these multiple notions of measurability as well). The notion of the integral we use is the ***Bochner integral***. This is well understood for Banach spaces [12] and is often mentioned in an offhand manner as being "the same" for locally convex spaces, e.g., [36, p. 96]. A detailed textbook treatment does not appear to exist, but fortunately this has been worked out in the note [5], to which we shall refer for details as needed. One has a notion of simple functions, meaning functions that are finite linear combinations, with coefficients in V, of characteristic functions of measurable sets. The ***integral*** of a simple function $\sigma = \sum_{j=1}^{k} v_j \chi_{A_j}$ is

$$\int_{\mathcal{T}} \sigma \, \mathrm{d}\mu = \sum_{j=1}^{k} \mu(A_j) v_j,$$

in the usual manner. A measurable function Ψ is ***Bochner approximable*** if it can be approximated with respect to any continuous seminorm by a net of simple functions. A Bochner approximable function Ψ is ***Bochner integrable*** if there is a net of simple functions approximating Ψ whose integrals converge in V to a unique value, which is called the ***integral*** of Ψ. If V is separable and complete, as will be the case for us in this work, then a measurable function $\Psi \colon \mathcal{T} \to \mathsf{V}$ is Bochner integrable if and only if

$$\int_{\mathcal{T}} p \circ \Psi \, \mathrm{d}\mu < \infty$$

---

[2] Most of the theories of integration in locally convex spaces coincide for the sorts of locally convex spaces we deal with.

for every continuous seminorm p on V [5, Theorems 3.2 and 3.3]. This construction of the integral clearly agrees with the standard construction of the Lebesgue integral for functions taking values in $\mathbb{R}$ or $\mathbb{C}$ (or any finite-dimensional vector space over $\mathbb{R}$ or $\mathbb{C}$, for that matter). If $A \subseteq \mathsf{V}$, by $\mathrm{L}^1(\mathcal{T}; A)$ we denote the space of Bochner integrable functions with values in A. The space $\mathrm{L}^1(\mathcal{T}; \mathsf{V})$ is itself a locally convex topological vector space with topology defined by the seminorms

$$\hat{p}(\Psi) = \int_{\mathcal{T}} p \circ \Psi \,\mathrm{d}\mu,$$

where p is a continuous seminorm for V [36, p. 96]. In the case where $\mathcal{T} = I$ is an interval in $\mathbb{R}$, $\mathrm{L}^1_{\mathrm{loc}}(I; A)$ denotes the set of locally integrable functions, i.e., those functions whose restriction to any compact subinterval is integrable.

While it does not generally make sense to talk about integrability of measurable functions with values in a topological space, one *can* sensibly talk about essentially bounded functions. This means that one needs a notion of boundedness, this being supplied by a bornology. A ***bornology*** on a set $\mathcal{S}$ is a family $\mathscr{B}$ of subsets of $\mathcal{S}$, called ***bounded sets***, and satisfying the axioms:

1. $\mathcal{S}$ is covered by bounded sets, i.e., $\mathcal{S} = \cup_{B \in \mathscr{B}} B$;
2. subsets of bounded sets are bounded, i.e., if $B \in \mathscr{B}$ and if $A \subseteq B$, then $A \in \mathscr{B}$;
3. finite unions of bounded sets are bounded, i.e., if $B_1, \dots, B_k \in \mathscr{B}$, then $\cup_{j=1}^k B_j \in \mathscr{B}$.

Bornologies are less popular than topologies, but a treatment in some generality can be found in [22]. There are two bornologies we consider in this monograph. One is the ***compact bornology*** for a topological space $\mathcal{X}$ whose bounded sets are the relatively compact sets. The other is the ***von Neumann bornology*** for a locally convex topological vector space V whose bounded sets are those subsets $\mathcal{B} \subseteq \mathsf{V}$ for which, for any neighbourhood $\mathcal{N}$ of $0 \in \mathsf{V}$, there exists $\lambda \in \mathbb{R}_{>0}$ such that $\mathcal{B} \subseteq \lambda\mathcal{N}$. On any locally convex topological vector space we thus have these two bornologies, and generally they are not the same. Indeed, if V is an infinite-dimensional normed vector space, then the compact bornology is strictly contained in the von Neumann bornology. We will, in fact, have occasion to use both of these bornologies and shall make it clear which we mean. Now, if $(\mathcal{T}, \mathscr{M}, \mu)$ is a measure space and if $(\mathcal{X}, \mathscr{B})$ is a bornological space, i.e., a set $\mathcal{X}$ with a bornology $\mathscr{B}$, a measurable map $\Psi : \mathcal{T} \to \mathcal{X}$ is ***essentially bounded*** if there exists a bounded set $B \subseteq \mathcal{X}$ such that

$$\mu(\{t \in \mathcal{T} \mid \Psi(t) \notin B\}) = 0.$$

By $\mathrm{L}^\infty(\mathcal{T}; \mathcal{X})$ we denote the set of essentially bounded maps. If $\mathcal{T} = I$ is an interval in $\mathbb{R}$, a measurable map $\Psi : I \to \mathcal{X}$ is ***locally essentially bounded*** in the bornology $\mathscr{B}$ if $\Psi|J$ is essentially bounded in the bornology $\mathscr{B}$ for every compact subinterval $J \subseteq I$. By $\mathrm{L}^\infty_{\mathrm{loc}}(I; \mathcal{X})$ we denote the set of locally essentially bounded maps; thus the bornology is to be understood when we write expressions such as this.

Intended Readership

It is our intention that this monograph be useful for a control theoretic audience, since one of the areas where this work is likely to have a substantial impact is in control theory. For this reason, we have written the results in a manner that is perhaps more detailed than usual, and with many more and more detailed references than usual. Much of the mathematical material we use in this work is not a part of the standard background of a control theoretic audience. This is especially true of our use of locally convex topologies and holomorphic geometry, and for these subjects we include various discussions, some as footnotes, that provide a little historical and mathematical context in these areas. By doing this, our hope is that the presentation is self-contained and comprehensive enough for a reader to get most of the ideas needed from the monograph and the references.

References

1. Abraham, R., Marsden, J.E., Ratiu, T.S.: Manifolds, Tensor Analysis, and Applications, 2 edn. No. 75 in Applied Mathematical Sciences. Springer-Verlag (1988)
2. Agrachev, A.A., Gamkrelidze, R.V.: The exponential representation of flows and the chronological calculus. Mathematics of the USSR-Sbornik **107**(4), 467–532 (1978)
3. Agrachev, A.A., Gamkrelidze, R.V.: Local controllability and semigroups of diffeomorphisms. Acta Applicandae Mathematicae. An International Journal on Applying Mathematics and Mathematical Applications **32**(1), 1–57 (1993)
4. Agrachev, A.A., Sachkov, Y.: Control Theory from the Geometric Viewpoint, *Encyclopedia of Mathematical Sciences*, vol. 87. Springer-Verlag, New York/Heidelberg/Berlin (2004)
5. Beckmann, R., Deitmar, A.: Strong vector valued integrals (2011). URL http://arxiv.org/abs/1102.1246v1. ArXiv:1102.1246v1 [math.FA]
6. Bourbaki, N.: Algebra I. Elements of Mathematics. Springer-Verlag, New York/Heidelberg/-Berlin (1989)
7. Bourbaki, N.: Algebra II. Elements of Mathematics. Springer-Verlag, New York/Heidelberg/-Berlin (1990)
8. Cieliebak, K., Eliashberg, Y.: From Stein to Weinstein and Back: Symplectic Geometry of Affine Complex Manifolds. No. 59 in American Mathematical Society Colloquium Publications. American Mathematical Society, Providence, RI (2012)
9. Coddington, E.E., Levinson, N.: Theory of Ordinary Differential Equations, 8 edn. Robert E. Krieger Publishing Company, Huntington/New York (1984)
10. Conway, J.B.: A Course in Functional Analysis, 2 edn. No. 96 in Graduate Texts in Mathematics. Springer-Verlag, New York/Heidelberg/Berlin (1985)
11. Demailly, J.P.: Complex analytic and differential geometry. Unpublished manuscript made publicly available (2012). URL http://www-fourier.ujf-grenoble.fr/~demailly/manuscripts/agbook.pdf
12. Diestel, J., Uhl, Jr., J.J.: Vector Measures. No. 15 in American Mathematical Society Mathematical Surveys and Monographs. American Mathematical Society, Providence, RI (1977)
13. Domański, P.: Notes on real analytic functions and classical operators. In: O. Blasco, J. Bonet, J. Calabuig, D. Jornet (eds.) Proceedings of the Third Winter School in Complex Analysis and Operator Theory, *Contemporary Mathematics*, vol. 561, pp. 3–47. American Mathematical Society, Providence, RI (2010)
14. Domański, P., Vogt, D.: The space of real-analytic functions has no basis. Polska Akademia Nauk. Instytut Matematyczny. Studia Mathematica **142**(2), 187–200 (2000)

15. Fritzsche, K., Grauert, H.: From Holomorphic Functions to Complex Manifolds. No. 213 in Graduate Texts in Mathematics. Springer-Verlag, New York/Heidelberg/Berlin (2002)
16. Fuller, A.T.: Relay control systems optimized for various performance criteria. In: Proceedings of the First IFAC World Congress, pp. 510–519. IFAC, Butterworth & Co., Ltd. London, Moscow (1960)
17. Groethendieck, A.: Topological Vector Spaces. Notes on Mathematics and its Applications. Gordon & Breach Science Publishers, New York (1973)
18. Gunning, R.C.: Introduction to Holomorphic Functions of Several Variables. Volume I: Function Theory. Wadsworth & Brooks/Cole Mathematics Series. Wadsworth & Brooks/Cole, Belmont, CA (1990)
19. Gunning, R.C.: Introduction to Holomorphic Functions of Several Variables. Volume II: Local Theory. Wadsworth & Brooks/Cole Mathematics Series. Wadsworth & Brooks/Cole, Belmont, CA (1990)
20. Gunning, R.C.: Introduction to Holomorphic Functions of Several Variables. Volume III: Homological Theory. Wadsworth & Brooks/Cole Mathematics Series. Wadsworth & Brooks/Cole, Belmont, CA (1990)
21. Gunning, R.C., Rossi, H.: Analytic Functions of Several Complex Variables. American Mathematical Society, Providence, RI (1965). 2009 reprint by AMS
22. Hogbe-Nlend, H.: Bornologies and Functional Analysis. No. 26 in North Holland Mathematical Studies. North-Holland, Amsterdam/New York (1977). Translated from the French by V. B. Moscatelli
23. Hogbe-Nlend, H., Moscatelli, V.B.: Nuclear and Conuclear Spaces. No. 52 in North Holland Mathematical Studies. North-Holland, Amsterdam/New York (1981)
24. Hörmander, L.: An Introduction to Complex Analysis in Several Variables, 2 edn. North-Holland, Amsterdam/New York (1973)
25. Horváth, J.: Topological Vector Spaces and Distributions. Vol. I. Addison Wesley, Reading, MA (1966)
26. Hungerford, T.W.: Algebra. No. 73 in Graduate Texts in Mathematics. Springer-Verlag, New York/Heidelberg/Berlin (1980)
27. Jarchow, H.: Locally Convex Spaces. Mathematical Textbooks. Teubner, Leipzig (1981)
28. Kawski, M.: Control variations with an increasing number of switchings. American Mathematical Society. Bulletin. New Series **18**(2), 149–152 (1988)
29. Kobayashi, S., Nomizu, K.: Foundations of Differential Geometry, Volume I. No. 15 in Interscience Tracts in Pure and Applied Mathematics. Interscience Publishers, New York (1963)
30. Kolář, I., Michor, P.W., Slovák, J.: Natural Operations in Differential Geometry. Springer-Verlag, New York/Heidelberg/Berlin (1993)
31. Krantz, S.G., Parks, H.R.: A Primer of Real Analytic Functions, 2 edn. Birkhäuser Advanced Texts. Birkhäuser, Boston/Basel/Stuttgart (2002)
32. Pietsch, A.: Nuclear Locally Convex Spaces. No. 66 in Ergebnisse der Mathematik und ihrer Grenzgebiete. Springer-Verlag, New York/Heidelberg/Berlin (1969)
33. Remmert, R.: Theorie der Modifikationen. I. Stetige und eigentliche Modifikationen komplexer Räume. Mathematische Annalen **129**, 274–296 (1955)
34. Rudin, W.: Functional Analysis, 2 edn. International Series in Pure and Applied Mathematics. McGraw-Hill, New York (1991)
35. Saunders, D.J.: The Geometry of Jet Bundles. No. 142 in London Mathematical Society Lecture Note Series. Cambridge University Press, New York/Port Chester/Melbourne/Sydney (1989)
36. Schaefer, H.H., Wolff, M.P.: Topological Vector Spaces, 2 edn. No. 3 in Graduate Texts in Mathematics. Springer-Verlag, New York/Heidelberg/Berlin (1999)
37. Schuricht, F., von der Mosel, H.: Ordinary differential equations with measurable right-hand side and parameter dependence. Tech. Rep. Preprint 676, Universität Bonn, SFB 256 (2000)
38. Sontag, E.D.: Mathematical Control Theory: Deterministic Finite Dimensional Systems, 2 edn. No. 6 in Texts in Applied Mathematics. Springer-Verlag, New York/Heidelberg/Berlin (1998)

39. Vogt, D.: A fundamental system of seminorms for $A(K)$ (2013). URL http://arxiv.org/abs/1309.6292v1. ArXiv:1309.6292v1 [math.FA]
40. Whitney, H.: Differentiable manifolds. Annals of Mathematics. Second Series **37**(3), 645–680 (1936)
41. Zelikin, M.I., Borisov, V.F.: Theory of Chattering Control. Systems & Control: Foundations & Applications. Birkhäuser, Boston/Basel/Stuttgart (1994)

Chapter 2
Fibre Metrics for Jet Bundles

One of the principal devices we use in the monograph is convenient seminorms for the various topologies we use for spaces of sections of vector bundles. Since such topologies rely on placing suitable norms on derivatives of sections, i.e., on jet bundles of vector bundles, in this chapter we present a means for defining such norms, using as our starting point a pair of connections, one for the base manifold, and one for the vector bundle. These allow us to provide a direct sum decomposition of the jet bundle into its component "derivatives", and so then a natural means of defining a fibre metric for jet bundles using metrics on the tangent bundle of the base manifold and the fibres of the vector bundle.

As we shall see, in the smooth case these constructions are a convenience, whereas in the real analytic case they provide a crucial ingredient in our global, coordinate-free description of seminorms for the topology of the space of real analytic sections of a vector bundle. For this reason, in this chapter we shall also consider the existence of, and some properties of, real analytic connections in vector bundles.

2.1 A Decomposition for the Jet Bundles of a Vector Bundle

We let $\pi\colon \mathsf{E} \to \mathsf{M}$ be a smooth vector bundle with $\pi_m\colon \mathsf{J}^m\mathsf{E} \to \mathsf{M}$ its mth jet bundle. In a local trivialisation of $\mathsf{J}^m\mathsf{E}$, the fibres of this vector bundle are

$$\oplus_{j=0}^{m} \mathrm{L}^{j}_{\mathrm{sym}}(\mathbb{R}^n;\mathbb{R}^k),$$

with n the dimension of M and k the fibre dimension of E. This decomposition of the derivatives, order-by-order, that we see in the local trivialisation has no global analogue, but such a decomposition can be provided with the use of connections, and we describe how to do this.

S. Jafarpour, A.D. Lewis, *Time-Varying Vector Fields and Their Flows*, SpringerBriefs in Mathematics, DOI 10.1007/978-3-319-10139-2_2

We suppose that we have a linear connection ∇^0 on the vector bundle E and an affine connection ∇ on M. We then have a connection, that we also denote by ∇, on $\mathsf{T}^*\mathsf{M}$ defined by

$$\mathscr{L}_Y\langle\alpha; X\rangle = \langle\nabla_Y\alpha; X\rangle + \langle\alpha; \nabla_Y X\rangle.$$

For $\xi \in \Gamma^\infty(\mathsf{E})$ we then have $\nabla^0\xi \in \Gamma^\infty(\mathsf{T}^*\mathsf{M} \otimes \mathsf{E})$ defined by $\nabla^0\xi(X) = \nabla^0_X\xi$ for $X \in \Gamma^\infty(\mathsf{TM})$. The connections ∇^0 and ∇ extend naturally to a connection, that we denote by ∇^m, on $\mathrm{T}^m(\mathsf{T}^*\mathsf{M}) \otimes \mathsf{E}$, $m \in \mathbb{Z}_{>0}$, by the requirement that

$$\nabla^m_X(\alpha^1 \otimes \cdots \otimes \alpha^m \otimes \xi)$$
$$= \sum_{j=1}^{m}(\alpha^1 \otimes \cdots \otimes (\nabla_X\alpha_j) \otimes \cdots \otimes \alpha^m \otimes \xi) + \alpha^1 \otimes \cdots \otimes \alpha^m \otimes (\nabla^0_X\xi)$$

for $\alpha^1, \dots, \alpha^m \in \Gamma^\infty(\mathsf{T}^*\mathsf{M})$ and $\xi \in \Gamma^\infty(\mathsf{E})$. Note that

$$\nabla^{(m)}\xi \triangleq \nabla^m(\nabla^{m-1}\cdots(\nabla^1(\nabla^0\xi))) \in \Gamma^\infty(\mathrm{T}^{m+1}(\mathsf{T}^*\mathsf{M}) \otimes \mathsf{E}). \tag{2.1}$$

Now, given $\xi \in \Gamma^\infty(\mathsf{E})$ and $m \in \mathbb{Z}_{\geq 0}$, we define

$$P^{m+1}_{\nabla,\nabla^0}(\xi) = \mathrm{Sym}_{m+1} \otimes \mathrm{id}_{\mathsf{E}}(\nabla^{(m)}\xi) \in \Gamma^\infty(\mathrm{S}^{m+1}(\mathsf{T}^*\mathsf{M}) \otimes \mathsf{E}),$$

where $\mathrm{Sym}_m \colon \mathrm{T}^m(\mathsf{V}) \to \mathrm{S}^m(\mathsf{V})$ is defined by

$$\mathrm{Sym}_m(v_1 \otimes \cdots \otimes v_m) = \frac{1}{m!}\sum_{\sigma\in\mathfrak{S}_m} v_{\sigma(1)} \otimes \cdots \otimes v_{\sigma(m)}.$$

We take the convention that $P^0_{\nabla,\nabla^0}(\xi) = \xi$.

The following lemma is then key for our presentation. While this lemma exists in the literature in various forms, often in the form of results concerning the extension of connections by "bundle functors", e.g., [7, Chap. X], we were unable to find the succinct statement we give here. Pohl [10] gives existential results dual to what we give here, but stops short of giving an explicit formula such as we give below. For this reason, we give a complete proof of the lemma.

Lemma 2.1 (Decomposition of Jet Bundles). *The map*

$$S^m_{\nabla,\nabla^0} \colon \mathsf{J}^m\mathsf{E} \to \oplus_{j=0}^m(\mathrm{S}^j(\mathsf{T}^*\mathsf{M}) \otimes \mathsf{E})$$
$$j_m\xi(x) \mapsto (\xi(x), P^1_{\nabla,\nabla^0}(\xi)(x), \dots, P^m_{\nabla,\nabla^0}(\xi)(x))$$

is an isomorphism of vector bundles, and, for each $m \in \mathbb{Z}_{>0}$, the diagram

$$\begin{array}{ccc} \mathsf{J}^{m+1}\mathsf{E} & \xrightarrow{S^{m+1}_{\nabla,\nabla^0}} & \oplus_{j=0}^{m+1}(\mathsf{S}^j(\mathsf{T}^*\mathsf{M}) \otimes \mathsf{E}) \\ {\scriptstyle \pi^{m+1}_m}\downarrow & & \downarrow{\scriptstyle \mathrm{pr}^{m+1}_m} \\ \mathsf{J}^m\mathsf{E} & \xrightarrow[S^m_{\nabla,\nabla^0}]{} & \oplus_{j=0}^{m}(\mathsf{S}^j(\mathsf{T}^*\mathsf{M}) \otimes \mathsf{E}) \end{array}$$

commutes, where pr^{m+1}_m is the obvious projection, stripping off the last component of the direct sum.

Proof: We prove the result by induction on m. For $m = 0$ the result is a tautology. For $m = 1$, as in [7, Sect. 17.1], we have a vector bundle mapping $S_{\nabla^0}: \mathsf{E} \to \mathsf{J}^1\mathsf{E}$ over id_M that determines the connection ∇^0 by

$$\nabla^0\xi(x) = j_1\xi(x) - S_{\nabla^0}(\xi(x)). \tag{2.2}$$

Let us show that S^1_{∇,∇^0} is well defined. Thus let $\xi, \eta \in \Gamma^\infty(\mathsf{E})$ be such that $j_1\xi(x) = j_1\eta(x)$. Then, clearly, $\xi(x) = \eta(x)$, and the formula (2.2) shows that $\nabla\xi(x) = \nabla\eta(x)$, and so S^1_{∇,∇^0} is indeed well defined. It is clearly linear on fibres, so it remains to show that it is an isomorphism. This will follow from dimension counting if it is injective. However, if $S^1_{\nabla,\nabla^0}(j_1\xi(x)) = 0$, then $j_1\xi(x) = 0$ by (2.2).

For the induction step, we begin with a sublemma.

Sublemma 1 *Let F be a field and consider the following commutative diagram of finite-dimensional F-vector spaces with exact rows and columns:*

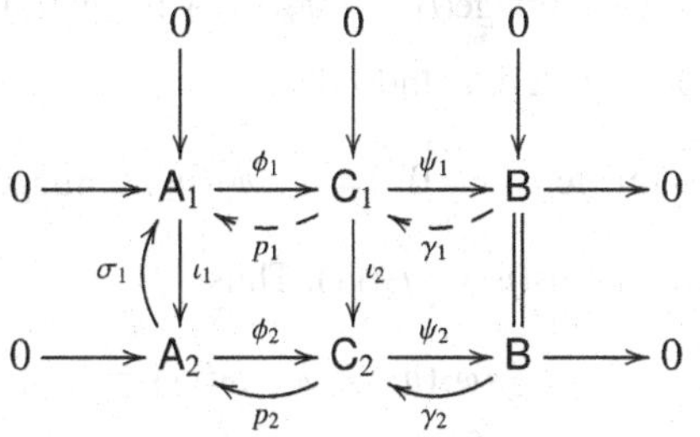

If there exists a mapping $\gamma_2 \in \mathrm{Hom}_\mathsf{F}(\mathsf{B};\mathsf{C}_2)$ such that $\psi_2 \circ \gamma_2 = \mathrm{id}_\mathsf{B}$ (with $p_2 \in \mathrm{Hom}_\mathsf{F}(\mathsf{C}_2;\mathsf{A}_2)$ the corresponding projection), then there exists a unique mapping $\gamma_1 \in \mathrm{Hom}_\mathsf{F}(\mathsf{B};\mathsf{C}_1)$ such that $\psi_1 \circ \gamma_1 = \mathrm{id}_\mathsf{B}$ and such that $\gamma_2 = \iota_2 \circ \gamma_1$. There is also induced a projection $p_1 \in \mathrm{Hom}_\mathsf{F}(\mathsf{C}_1;\mathsf{A}_1)$.

Moreover, if there additionally exists a mapping $\sigma_1 \in \mathrm{Hom}_\mathsf{F}(\mathsf{A}_2;\mathsf{A}_1)$ such that $\sigma_1 \circ \iota_1 = \mathrm{id}_{\mathsf{A}_1}$, then the projection p_1 is uniquely determined by the condition $p_1 = \sigma_1 \circ p_2 \circ \iota_2$.

Proof: We begin by extending the diagram to one of the form

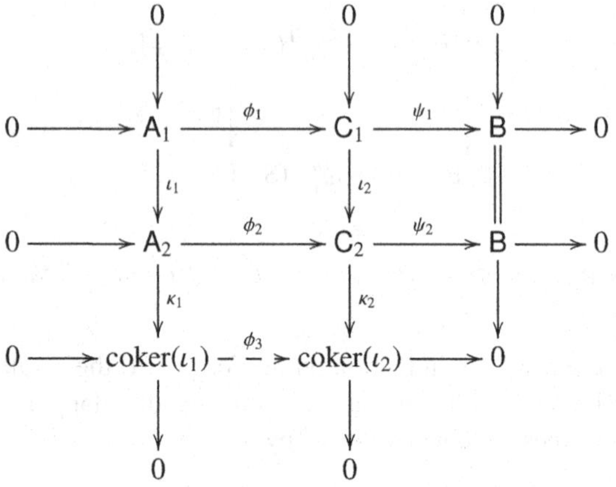

also with exact rows and columns. We claim that there is a natural mapping ϕ_3 between the cokernels, as indicated by the dashed arrow in the diagram, and that ϕ_3 is, moreover, an isomorphism. Suppose that $u_2 \in \text{image}(\iota_1)$ and let $u_1 \in \mathsf{A}_1$ be such that $\iota_1(u_1) = u_2$. By commutativity of the diagram, we have

$$\phi_2(u_2) = \phi_2 \circ \iota_1(u_1) = \iota_2 \circ \phi_1(u_1),$$

showing that $\phi_2(\text{image}(\iota_1)) \subseteq \text{image}(\iota_2)$. We thus have a well-defined homomorphism

$$\begin{aligned} \phi_3\colon \text{coker}(\iota_1) &\to \text{coker}(\iota_2) \\ u_2 + \text{image}(\iota_1) &\mapsto \phi_2(u_2) + \text{image}(\iota_2). \end{aligned}$$

We now claim that ϕ_3 is injective. Indeed,

$$\phi_3(u_2 + \text{image}(\iota_1)) = 0 \quad \Longrightarrow \quad \phi_2(u_2) \in \text{image}(\iota_2).$$

Thus let $v_1 \in \mathsf{C}_1$ be such that $\phi_2(u_2) = \iota_2(v_1)$. Thus

$$\begin{aligned} 0 = \psi_2 \circ \phi_2(u_2) = \psi_2 \circ \iota_2(v_1) = \psi_1(v_1) \\ \Longrightarrow \quad v_1 \in \ker(\psi_1) = \text{image}(\phi_1). \end{aligned}$$

Thus $v_1 = \phi_1(u_1')$ for some $u_1' \in \mathsf{A}_1$. Therefore,

$$\phi_2(u_2) = \iota_2 \circ \phi_1(u_1') = \phi_2 \circ \iota_1(u_1'),$$

and injectivity of ϕ_2 gives $u_2 \in \text{image}(\iota_1)$ and so $u_2 + \text{image}(\iota_1) = 0 + \text{image}(\iota_1)$, giving the desired injectivity of ϕ_3.

Now note that

$$\dim(\text{coker}(\iota_1)) = \dim(\mathsf{A}_2) - \dim(\mathsf{A}_1)$$

by exactness of the left column. Also,

$$\dim(\mathrm{coker}(\iota_2)) = \dim(\mathsf{C}_2) - \dim(\mathsf{C}_1)$$

by exactness of the middle column. By exactness of the top and middle rows, we have

$$\dim(\mathsf{B}) = \dim(\mathsf{C}_2) - \dim(\mathsf{A}_2) = \dim(\mathsf{C}_1) - \dim(\mathsf{A}_1).$$

This proves that

$$\dim(\mathrm{coker}(\iota_1)) = \dim(\mathrm{coker}(\iota_2)).$$

Thus the homomorphism ϕ_3 is an isomorphism, as claimed.

Now we proceed with the proof, using the extended diagram, and identifying the bottom cokernels with the isomorphism ϕ_3. The existence of the stated homomorphism γ_2 means that the middle row in the diagram splits. Therefore, $\mathsf{C}_2 = \mathrm{image}(\phi_2) \oplus \mathrm{image}(\gamma_2)$. Thus there exists a well-defined projection $p_2 \in \mathrm{Hom}_{\mathsf{F}}(\mathsf{C}_2; \mathsf{A}_2)$ such that $p_2 \circ \phi_2 = \mathrm{id}_{\mathsf{A}_2}$ [5, Theorem 41.1].

We will now prove that $\mathrm{image}(\gamma_2) \subseteq \mathrm{image}(\iota_2)$. By commutativity of the diagram and since ψ_1 is surjective, if $w \in \mathsf{B}$, then there exists $v_1 \in \mathsf{C}_1$ such that $\psi_2 \circ \iota_2(v_1) = w$. Since $\psi_2 \circ \gamma_2 = \mathrm{id}_{\mathsf{B}}$, we have

$$\psi_2 \circ \iota_2(v_1) = \psi_2 \circ \gamma_2(w) \quad \Longrightarrow \quad \iota_2(v_1) - \gamma_2(w) \in \ker(\psi_2) = \mathrm{image}(\phi_2).$$

Let $u_2 \in \mathsf{A}_2$ be such that $\phi_2(u_2) = \iota_2(v_1) - \gamma_2(w)$. Since $p_2 \circ \phi_2 = \mathrm{id}_{\mathsf{A}_2}$ we have

$$u_2 = p_2 \circ \iota_2(v_1) - p_2 \circ \gamma_2(w),$$

whence

$$\kappa_1(u_2) = \kappa_1 \circ p_2 \circ \iota_2(v_1) - \kappa_1 \circ p_2 \circ \gamma_2(w) = 0,$$

noting that (1) $\kappa_1 \circ p_2 = \kappa_2$ (by commutativity), (2) $\kappa_2 \circ \iota_2 = 0$ (by exactness), and (3) $p_2 \circ \gamma_2 = 0$ (by exactness). Thus $u_2 \in \ker(\kappa_1) = \mathrm{image}(\iota_1)$. Let $u_1 \in \mathsf{A}_1$ be such that $\iota_1(u_1) = u_2$. We then have

$$\iota_2(v_1) - \gamma_2(w) = \phi_2 \circ \iota_1(u_1) = \iota_2 \circ \phi_1(u_1),$$

which gives $\gamma_2(w) \in \mathrm{image}(\iota_2)$, as claimed.

Now we define $\gamma_1 \in \mathrm{Hom}_{\mathsf{F}}(\mathsf{B}; \mathsf{C}_1)$ by asking that $\gamma_1(w) \in \mathsf{C}_1$ have the property that $\iota_2 \circ \gamma_1(w) = \gamma_2(w)$, this making sense since we just showed that $\mathrm{image}(\gamma_2) \subseteq \mathrm{image}(\iota_2)$. Moreover, since ι_2 is injective, the definition uniquely prescribes γ_1. Finally we note that

$$\psi_1 \circ \gamma_1 = \psi_2 \circ \iota_2 \circ \gamma_1 = \psi_2 \circ \gamma_2 = \mathrm{id}_{\mathsf{B}},$$

as claimed.

To prove the final assertion, let us denote $\hat{p}_1 = \sigma_1 \circ p_2 \circ \iota_2$. We then have

$$\hat{p}_1 \circ \phi_1 = \sigma_1 \circ p_2 \circ \iota_2 \circ \phi_1 = \sigma_1 \circ p_2 \circ \phi_2 \circ \iota_1 = \sigma_1 \circ \iota_1 = \mathrm{id}_{\mathsf{A}_1},$$

using commutativity. We also have

$$\hat{p}_1 \circ \gamma_1 = \sigma_1 \circ p_2 \circ \iota_2 \circ \gamma_1 = \sigma_1 \circ p_2 \circ \gamma_2 = 0.$$

The two preceding conclusions show that $\hat{p}_1$ is the projection defined by the splitting of the top row of the diagram, i.e., $\hat{p}_1 = p_1$. ▽

Now suppose that the lemma is true for $m \in \mathbb{Z}_{>0}$. For any $k \in \mathbb{Z}_{>0}$ we have a short exact sequence

$$0 \longrightarrow \mathsf{S}^k(\mathsf{T}^*\mathsf{M}) \otimes \mathsf{E} \xrightarrow{\epsilon_k} \mathsf{J}^k\mathsf{E} \xrightarrow{\pi^k_{k-1}} \mathsf{J}^{k-1}\mathsf{E} \longrightarrow 0$$

for which we refer to [11, Theorem 6.2.9]. Recall from [11, Definition 6.2.25] that we have an inclusion $\iota_{1,m}$ of $\mathsf{J}^{m+1}\mathsf{E}$ in $\mathsf{J}^1(\mathsf{J}^m\mathsf{E})$ by $j_{m+1}\xi(x) \mapsto j_1(j_m\xi(x))$. We also have an induced injection

$$\hat{\iota}_{1,m} \colon \mathsf{S}^{m+1}(\mathsf{T}^*\mathsf{M}) \otimes \mathsf{E} \to \mathsf{T}^*\mathsf{M} \otimes \mathsf{J}^m\mathsf{E}$$

defined by the composition

$$\mathsf{S}^{m+1}(\mathsf{T}^*\mathsf{M}) \otimes \mathsf{E} \longrightarrow \mathsf{T}^*\mathsf{M} \otimes \mathsf{S}^m(\mathsf{T}^*\mathsf{M}) \otimes \mathsf{E} \xrightarrow{\mathrm{id} \otimes \epsilon_m} \mathsf{T}^*\mathsf{M} \otimes \mathsf{J}^m\mathsf{E}$$

Explicitly, the left arrow is defined by

$$\alpha^1 \odot \cdots \odot \alpha^{m+1} \otimes \xi \mapsto \sum_{j=1}^{m+1} \alpha^j \otimes \alpha^1 \odot \cdots \odot \alpha^{j-1} \odot \alpha^{j+1} \odot \cdots \odot \alpha^{m+1} \otimes \xi,$$

$\odot$ denoting the symmetric tensor product defined by

$$A \odot B = \sum_{\sigma \in \mathfrak{S}_{k,l}} \sigma(A \otimes B), \tag{2.3}$$

for $A \in \mathsf{S}^k(\mathsf{V})$ and $B \in \mathsf{S}^l(\mathsf{V})$, and with $\mathfrak{S}_{k,l}$ the subset of $\mathfrak{S}_{k+l}$ consisting of permutations σ satisfying

$$\sigma(1) < \cdots < \sigma(k), \quad \sigma(k+1) < \cdots < \sigma(k+l).$$

We thus have the following commutative diagram with exact rows and columns:

$$\begin{array}{ccccccccc}
 & & 0 & & 0 & & 0 & & \\
 & & \downarrow & & \downarrow & & \downarrow & & \\
0 & \longrightarrow & \mathsf{S}^{m+1}(\mathsf{T}^*\mathsf{M}) \otimes \mathsf{E} & \underset{P_{m+1}}{\overset{\epsilon_{m+1}}{\rightleftarrows}} & \mathsf{J}^{m+1}\mathsf{E} & \underset{\Gamma_{m+1}}{\overset{\pi^{m+1}_m}{\rightleftarrows}} & \mathsf{J}^m\mathsf{E} & \longrightarrow & 0 \\
 & & \lambda_{1,m} \uparrow\downarrow \hat{\iota}_{1,m} & & \downarrow \iota_{1,m} & & \| & & \\
0 & \longrightarrow & \mathsf{T}^*\mathsf{M} \otimes \mathsf{J}^m\mathsf{E} & \underset{P_{1,m}}{\overset{\epsilon_{1,m}}{\rightleftarrows}} & \mathsf{J}^1(\mathsf{J}^m\mathsf{E}) & \underset{\Gamma_{1,m}}{\overset{(\pi_m)_1}{\rightleftarrows}} & \mathsf{J}^m\mathsf{E} & \longrightarrow & 0
\end{array} \tag{2.4}$$

We shall define a connection on $(\pi_m)_1 \colon \mathsf{J}^1(\mathsf{J}^m\mathsf{E}) \to \mathsf{J}^m\mathsf{E}$ which gives a splitting $\Gamma_{1,m}$ and $P_{1,m}$ of the lower row in the diagram. By the sublemma, this will give a splitting Γ_{m+1} and P_{m+1} of the upper row, and so give a projection from $\mathsf{J}^{m+1}\mathsf{E}$ onto $\mathsf{S}^{m+1}(\mathsf{T}^*\mathsf{M}) \otimes \mathsf{E}$, which will allow us to prove the induction step. To compute P_{m+1} from the sublemma, we shall also give a map $\lambda_{1,m}$ as in the diagram so that $\lambda_{1,m} \circ \hat{\iota}_{1,m}$ is the identity on $\mathsf{S}^{m+1}(\mathsf{T}^*\mathsf{M}) \otimes \mathsf{E}$.

We start, under the induction hypothesis, by making the identification

$$\mathsf{J}^m\mathsf{E} \simeq \oplus_{j=0}^m \mathsf{S}^j(\mathsf{T}^*\mathsf{M}) \otimes \mathsf{E},$$

and consequently writing a section of $\mathsf{J}^m\mathsf{E}$ as

$$x \mapsto (\xi(x), P^1_{\nabla,\nabla^0}(\xi(x)), \dots, P^m_{\nabla,\nabla^0}(\xi(x))).$$

We then have a connection $\overline{\nabla}^m$ on $\mathsf{J}^m\mathsf{E}$ given by

$$\overline{\nabla}^m_X(\xi, P^1_{\nabla,\nabla^0}(\xi), \dots, P^m_{\nabla,\nabla^0}(\xi)) = (\nabla^0_X\xi, \nabla^1_X P^1_{\nabla,\nabla^0}(\xi), \dots, \nabla^m_X P^m_{\nabla,\nabla^0}(\xi)).$$

Thus

$$\overline{\nabla}^m(\xi, P^1_{\nabla,\nabla^0}(\xi), \dots, P^m_{\nabla,\nabla^0}(\xi)) = (\nabla^0\xi, \nabla^1 P^1_{\nabla,\nabla^0}(\xi), \dots, \nabla^m P^m_{\nabla,\nabla^0}(\xi)),$$

which—according to the jet bundle characterisation of connections from [7, Sect. 17.1] and which we have already employed in (2.2)—gives the mapping $P_{1,m}$ in the diagram (2.4) as

$$P_{1,m}(j_1(\xi, P^1_{\nabla,\nabla^0}(\xi), \dots, P^m_{\nabla,\nabla^0}(\xi))) = (\nabla^0\xi, \nabla^1 P^1_{\nabla,\nabla^0}(\xi), \dots, \nabla^m P^m_{\nabla,\nabla^0}(\xi)).$$

Now we define a mapping $\lambda_{1,m}$ for which $\lambda_{1,m} \circ \hat{\iota}_{1,m}$ is the identity on $\mathsf{S}^{m+1}(\mathsf{T}^*\mathsf{M}) \otimes \mathsf{E}$. We continue to use the induction hypothesis in writing elements of $\mathsf{J}^m\mathsf{E}$, so that we consider elements of $\mathsf{T}^*\mathsf{M} \otimes \mathsf{J}^m\mathsf{E}$ of the form

$$(\alpha \otimes \xi, \alpha \otimes A_1, \dots, \alpha \otimes A_m),$$

for $\alpha \in \mathsf{T}^*\mathsf{M}$ and $A_k \in \mathsf{S}^k(\mathsf{T}^*\mathsf{M}) \otimes \mathsf{E}$, $k \in \{1, \dots, m\}$. We then define $\lambda_{1,m}$ by

$$\begin{aligned}\lambda_{1,m}(\alpha_0 \otimes \xi, \alpha_0 \otimes \alpha_1^1 \otimes \xi, \dots, \alpha_0 \otimes \alpha_m^1 \odot \cdots \odot \alpha_m^m \otimes \xi) \\ = \mathrm{Sym}_{m+1} \otimes \mathrm{id}_{\mathsf{E}}(\alpha_0 \otimes \alpha_m^1 \odot \cdots \odot \alpha_m^m \otimes \xi).\end{aligned}$$

Note that, with the form of $\mathsf{J}^m\mathsf{E}$ from the induction hypothesis, we have

$$\begin{aligned}&\hat{\iota}_{1,m}(\alpha^1 \odot \cdots \odot \alpha^{m+1} \otimes \xi) \\ &\quad = \Big(0, \dots, 0, \frac{1}{m+1} \sum_{j=1}^{m+1} \alpha^j \otimes \alpha^1 \odot \cdots \odot \alpha^{j-1} \odot \alpha^{j+1} \odot \cdots \odot \alpha^{m+1} \otimes \xi\Big).\end{aligned}$$

We then directly verify that $\lambda_{1,m} \circ \hat{\iota}_{1,m}$ is indeed the identity.

We finally claim that

$$P_{m+1}(j_{m+1}\xi(x)) = P^{m+1}_{\nabla,\nabla^0}(\xi), \tag{2.5}$$

which will establish the lemma. To see this, first note that it suffices to define P_{m+1} on image(ϵ_{m+1}) since

1. $\mathsf{J}^{m+1}\mathsf{E} \simeq (\mathsf{S}^{m+1}(\mathsf{T}^*\mathsf{M}) \otimes \mathsf{E}) \oplus \mathsf{J}^m\mathsf{E}$,
2. P_{m+1} is zero on $\mathsf{J}^m\mathsf{E} \subseteq \mathsf{J}^{m+1}\mathsf{E}$ (thinking of the inclusion arising from the connection-induced isomorphism from the preceding item), and
3. $P_{m+1} \circ \epsilon_{m+1}$ is the identity map on $\mathsf{S}^{m+1}(\mathsf{T}^*\mathsf{M}) \otimes \mathsf{E}$.

In order to connect the algebra and the geometry, let us write elements of $\mathsf{S}^{m+1}(\mathsf{T}^*\mathsf{M})\otimes\mathsf{E}$ in a particular way. We let $x \in \mathsf{M}$ and let $f^1, \dots, f^{m+1}$ be smooth functions contained in the maximal ideal of $\mathrm{C}^\infty(\mathsf{M})$ at x, i.e., $f^j(x) = 0$, $j \in \{1, \dots, m+1\}$. Let ξ be a smooth section of E. We then can work with elements of $\mathsf{S}^{m+1}(\mathsf{T}^*\mathsf{M}) \otimes \mathsf{E}$ of the form

$$\mathrm{d}f^1(x) \odot \cdots \odot \mathrm{d}f^{m+1}(x) \otimes \xi(x).$$

We then have

$$\epsilon_{m+1}(\mathrm{d}f^1(x) \odot \cdots \odot \mathrm{d}f^{m+1}(x) \otimes \xi(x)) = j_{m+1}(f^1 \cdots f^{m+1}\xi)(x);$$

this is easy to see using the Leibniz Rule, cf. [3, Lemma 2.1]. (See [1, Supplement 2.4A] for a description of the higher-order Leibniz Rule.) Now, using the last part of the sublemma, we compute

$$\begin{aligned}
&P_{m+1}(j_{m+1}(f^1 \cdots f^{m+1}\xi)(x)) \\
&\quad = \lambda_{1,m} \circ P_{1,m} \circ \iota_{1,m}(j_{m+1}(f^1 \cdots f^{m+1}\xi)(x)) \\
&\quad = \lambda_{1,m} \circ P_{1,m}(j_1(f^1 \cdots f^{m+1}\xi, P^1_{\nabla,\nabla^0}(f^1 \cdots f^{m+1}\xi), \\
&\qquad \dots, P^m_{\nabla,\nabla^0}(f^1 \cdots f^{m+1}\xi))(x)) \\
&\quad = \lambda_{1,m}(\nabla^0(f^1 \cdots f^{m+1}\xi)(x), \nabla^1 P^1_{\nabla,\nabla^0}(f^1 \cdots f^{m+1}\xi)(x), \\
&\qquad \dots, \nabla^m P^m_{\nabla,\nabla^0}(f^1 \cdots f^{m+1}\xi)(x)) \\
&\quad = \mathrm{Sym}_{m+1} \otimes \mathrm{id}_{\mathsf{E}}(\nabla^m P^m_{\nabla,\nabla^0}(f^1 \cdots f^{m+1}\xi)(x)) \\
&\quad = P^{m+1}_{\nabla,\nabla_0}(f^1 \cdots f^{m+1}\xi)(x),
\end{aligned}$$

which shows that, with P_{m+1} defined as in (2.5), $P_{m+1} \circ \epsilon_{m+1}$ is indeed the identity on $\mathsf{S}^{m+1}(\mathsf{T}^*\mathsf{M}) \otimes \mathsf{E}$.

The commuting of the diagram in the statement of the lemma follows directly from the recursive nature of the constructions. □

Next we consider the decomposition of the preceding lemma in geodesic normal coordinates about a fixed point in M.

Lemma 2.2 (Jet Bundle Decomposition and Geodesic Normal Coordinates). *Let $r \in \{\infty, \omega\}$, let $\pi\colon \mathsf{E} \to \mathsf{M}$ be a C^r-vector bundle, let ∇^0 be a C^r-linear connection on E, and let ∇ be a C^r-affine connection on M. Let $x \in \mathsf{M}$, and let $\mathcal{N} \subseteq \mathsf{T}_x\mathsf{M}$ be a convex* *neighbourhood of* 0_x *and* $\mathcal{V} \subseteq \mathsf{M}$ *be a neighbourhood of* x *such that the exponential map* $\exp_x$ *corresponding to* ∇ *is a* C^r*-diffeomorphism from* $\mathcal{N}$ *to* $\mathcal{V}$. *For* $y \in \mathcal{V}$, *let* $\tau_{xy} \colon \mathsf{E}_x \to \mathsf{E}_y$ *be parallel transport along the geodesic* $t \mapsto \exp_x(t \exp_x^{-1}(y))$. *Define*

$$\begin{aligned} \kappa_x \colon \mathcal{N} \times \mathsf{E}_x &\to \mathsf{E}|\mathcal{V} \\ (v, e_x) &\mapsto \tau_{x,\exp_x(v)}(e_x). \end{aligned}$$

Then
(a) κ_x *is a* C^r*-vector bundle isomorphism over* $\exp_x$ *and*
(b) if $\xi \in \Gamma^r(\mathsf{E}|\mathcal{V})$, *then* $P^m_{\nabla,\nabla^0}(\xi)(x) = \boldsymbol{D}^m\hat{\xi}(0)$, *where* $\hat{\xi}$ *is the section of* $\mathcal{N} \times \mathsf{E}_x$ *defined by* $\hat{\xi}(v) = \tau^{-1}_{x,\exp_x(v)}(\xi(\exp_x(v)))$.

Proof: (a) Consider the vector field X_{∇,∇^0} on the Whitney sum $\mathsf{TM} \oplus \mathsf{E}$ defined by

$$X_{\nabla,\nabla^0}(v_x, e_x) = \mathrm{hlft}(v_x, v_x) \oplus \mathrm{hlft}_0(e_x, v_x),$$

where $\mathrm{hlft}(v_x, u_x)$ is the horizontal lift of $u_x \in \mathsf{T}_x\mathsf{M}$ to $\mathsf{T}_{v_x}\mathsf{TM}$ and $\mathrm{hlft}_0(e_x, u_x)$ is the horizontal lift of $u_x \in \mathsf{T}_x\mathsf{M}$ to $\mathsf{T}_{e_x}\mathsf{E}$. Note that, since

$$T\pi_{\mathsf{TM}}(\mathrm{hlft}(v_x, v_x)) = T\pi(\mathrm{hlft}_0(e_x, v_x)),$$

this is indeed a vector field on $\mathsf{TM} \oplus \mathsf{E}$. Moreover, the integral curve of X_{∇,∇^0} through (v_x, e_x) is $t \mapsto \gamma'(t) \oplus \tau(t)$, where γ is the geodesic with initial condition $\gamma'(0) = v_x$ and where $t \mapsto \tau(t)$ is parallel transport of e_x along γ. Thus X_{∇,∇^0} is a C^r-vector field, and so its flow depends in a C^r-manner on initial condition; [1, Lemma 4.1.9] for $r = \infty$ and [12, Proposition C.3.12] for $r = \omega$. In particular, it depends in a C^r-manner on initial conditions lying in $\mathcal{N} \times \mathsf{E}_x$. But, in this case, the map from initial condition to value at $t = 1$ is exactly κ_x. This shows that κ_x is indeed of class C^r. Moreover, it is clearly fibre preserving over $\exp_x$ and is linear on fibres, and so is a vector bundle map, cf. [1, Proposition 3.4.12(iii)].

(b) For $v \in \mathcal{N}$, let γ_v be the geodesic satisfying $\gamma_v'(0) = v$. Then, for $t \in \mathbb{R}_{>0}$ satisfying $|t| \leq 1$, define

$$\alpha_v(t) = \kappa_x^{-1} \circ \xi(\gamma_v(t)) = \tau^{-1}_{x,\gamma_v(t)}\xi(\gamma_v(t)).$$

We compute derivatives of α_v as follows, by induction and using the fact that $\nabla_{\gamma_v'(t)}\gamma_v'(t) = 0$:

$$\begin{aligned} \boldsymbol{D}\alpha_v(t) &= \tau^{-1}_{x,\gamma_v(t)}(\nabla^0\xi(\gamma_v'(t))) \\ \boldsymbol{D}^2\alpha_v(t) &= \tau^{-1}_{x,\gamma_v(t)}(\nabla^{(1)}\xi(\gamma_v'(t), \gamma_v'(t))) \\ &\vdots \\ \boldsymbol{D}^m\alpha_v(t) &= \tau^{-1}_{x,\gamma_v(t)}(\nabla^{(m-1)}\xi(\underbrace{\gamma_v'(t), \dots, \gamma_v'(t)}_{m \text{ times}})). \end{aligned}$$

Since $\gamma_v(t) = \gamma_{tv}(1)$ and since $\hat{\xi}(tv) = \alpha_v(t)$, we have

$$\boldsymbol{D}^m\alpha_v(0) = \boldsymbol{D}^m\hat{\xi}(0)(\underbrace{v, \dots, v}_{m \text{ times}}).$$

Thus

$$\boldsymbol{D}^m\hat{\xi}(0)(\underbrace{v, \dots, v}_{m \text{ times}}) = \nabla^{(m-1)}\xi(\underbrace{v, \dots, v}_{m \text{ times}}),$$

and the result follows by symmetrising both sides of this last equation. □

2.2 Fibre Metrics Using Jet Bundle Decompositions

We require the following result concerning inner products on tensor products.

Lemma 2.3 (Inner Products on Tensor Products). *Let* U *and* V *be finite-dimensional* $\mathbb{R}$*-vector spaces and let* $\mathbb{G}$ *and* $\mathbb{H}$ *be inner products on* U *and* V*, respectively. Then the element* $\mathbb{G} \otimes \mathbb{H}$ *of* $\mathrm{T}^2(\mathsf{U}^* \otimes \mathsf{V}^*)$ *defined by*

$$\mathbb{G} \otimes \mathbb{H}(u_1 \otimes v_1, u_2 \otimes v_2) = \mathbb{G}(u_1, u_2)\mathbb{H}(v_1, v_2)$$

is an inner product on $\mathsf{U} \otimes \mathsf{V}$.

Proof: Let $(e_1, \dots, e_m)$ and $(f_1, \dots, f_n)$ be orthonormal bases for U and V, respectively. Then

$$\{e_a \otimes f_j \mid a \in \{1, \dots, m\},\ j \in \{1, \dots, n\}\} \tag{2.6}$$

is a basis for $\mathsf{U} \otimes \mathsf{V}$. Note that

$$\mathbb{G} \otimes \mathbb{H}(e_a \otimes f_j, e_b \otimes f_k) = \mathbb{G}(e_a, e_b)\mathbb{H}(f_j, f_k) = \delta_{ab}\delta_{jk},$$

which shows that $\mathbb{G} \otimes \mathbb{H}$ is indeed an inner product, as (2.6) is an orthonormal basis. □

Now, we let $\mathbb{G}_0$ be a fibre metric on E and let $\mathbb{G}$ be a Riemannian metric on M. Let us denote by $\mathbb{G}^{-1}$ the associated fibre metric on $\mathsf{T}^*\mathsf{M}$ defined by

$$\mathbb{G}^{-1}(\alpha_x, \beta_x) = \mathbb{G}(\mathbb{G}^\sharp(\alpha_x), \mathbb{G}^\sharp(\beta_x)).$$

By induction using the preceding lemma, for each $j \in \mathbb{Z}_{>0}$ we have a fibre metric $\mathbb{G}_j$ on $\mathrm{T}^j(\mathsf{T}^*\mathsf{M}) \otimes \mathsf{E}$ induced by $\mathbb{G}^{-1}$ and $\mathbb{G}_0$. By restriction, this gives a fibre metric on $\mathrm{S}^j(\mathsf{T}^*\mathsf{M}) \otimes \mathsf{E}$. We can thus define a fibre metric $\overline{\mathbb{G}}_m$ on $\mathsf{J}^m\mathsf{E}$ given by

$$\overline{\mathbb{G}}_m(j_m\xi(x), j_m\eta(x)) = \sum_{j=0}^{m} \mathbb{G}_j\Big(\frac{1}{j!}P^j_{\nabla,\nabla^0}(\xi)(x), \frac{1}{j!}P^j_{\nabla,\nabla^0}(\eta)(x)\Big),$$

with the convention that $\nabla^{(-1)}\xi = \xi$. Associated with this inner product on fibres is the norm on fibres, which we denote by $\|\cdot\|_{\overline{\mathbb{G}}_m}$. We shall use these fibre norms continually in our descriptions of our various topologies below.

2.3 Real Analytic Connections

The fibre metrics from the preceding section will be used to define seminorms for spaces of sections of vector bundles. In the finitely differentiable and smooth cases, the particular fibre metrics we define above are not really required to give seminorms for the associated topologies: any fibre metrics on the jet bundles will suffice. Indeed, as long as one is only working with finitely many derivatives at one time, the choice of fibre norms on jet bundles is of no consequence, since different choices will be equivalent on compact subsets of M, cf. Sect. 3.1. However, when we work with the real analytic topology, we are no longer working only with finitely many derivatives, but with the infinite jet of a section. For this reason, different choices of fibre metric for jet bundles may give rise to different topologies for the space of real analytic sections, unless the behaviour of the fibre metrics is compatible as the order of derivatives goes to infinity. In this section we give a fundamental inequality for our fibre metrics of Sect. 2.2 in the real analytic case that ensures that they, in fact, describe the real analytic topology.

First let us deal with the matter of existence of real analytic data defining these fibre metrics.

Lemma 2.4 (Existence of Real Analytic Connections and Fibre Metrics). *If* $\pi\colon \mathsf{E} \to \mathsf{M}$ *is a real analytic vector bundle, then there exist*

(i) a real analytic linear connection on E,
(ii) a real analytic affine connection on M,
(iii) a real analytic fibre metric on E, *and*
(iv) a real analytic Riemannian metric on M.

Proof: By [4, Theorem 3], there exists a proper real analytic embedding ι_{E} of E in $\mathbb{R}^N$ for some $N \in \mathbb{Z}_{>0}$. There is then an induced proper real analytic embedding ι_{M} of M in $\mathbb{R}^N$ by restricting ι_{E} to the zero section of E. Let us take the subbundle $\hat{\mathsf{E}}$ of $\mathsf{T}\mathbb{R}^N|\iota_{\mathsf{M}}(\mathsf{M})$ whose fibre at $\iota_{\mathsf{M}}(x) \in \iota_{\mathsf{M}}(\mathsf{M})$ is

$$\hat{\mathsf{E}}_{\iota_{\mathsf{M}}(x)} = T_{0_x}\iota_{\mathsf{E}}(\mathsf{V}_{0_x}\mathsf{E}).$$

Now recall that $\mathsf{E} \simeq \zeta^*\mathsf{VE}$, where $\zeta\colon \mathsf{M} \to \mathsf{E}$ is the zero section [7, p. 55]. Let us abbreviate $\hat{\iota}_{\mathsf{E}} = T\iota_{\mathsf{E}}|\zeta^*\mathsf{VE}$. We then have the following diagram

$$\begin{array}{ccc} \mathsf{E} \simeq \zeta^*\mathsf{VE} & \xrightarrow{\hat{\iota}_{\mathsf{E}}} & \mathbb{R}^N \times \mathbb{R}^N \\ \downarrow \pi & & \downarrow \mathrm{pr}_2 \\ \mathsf{M} & \xrightarrow[\iota_{\mathsf{M}}]{} & \mathbb{R}^N \end{array} \qquad (2.7)$$

describing a monomorphism of real analytic vector bundles over the proper embedding ι_{M}, with the image of $\hat{\iota}_{\mathsf{E}}$ being $\hat{\mathsf{E}}$.

Among the many ways to prescribe a linear connection on the vector bundle E, we will take the prescription whereby one defines a mapping $K\colon \mathsf{TE} \to \mathsf{E}$ such that the two diagrams

$$\begin{array}{ccc} \mathsf{TE} & \xrightarrow{K} & \mathsf{E} \\ {\scriptstyle T\pi}\downarrow & & \downarrow{\scriptstyle \pi} \\ \mathsf{TM} & \xrightarrow[\pi_{\mathsf{TM}}]{} & \mathsf{M} \end{array} \qquad \begin{array}{ccc} \mathsf{TE} & \xrightarrow{K} & \mathsf{E} \\ {\scriptstyle \pi_{\mathsf{TE}}}\downarrow & & \downarrow{\scriptstyle \pi} \\ \mathsf{E} & \xrightarrow[\pi]{} & \mathsf{M} \end{array} \tag{2.8}$$

define vector bundle mappings [7, Sect. 11.11]. We define K as follows. For $e_x \in \mathsf{E}_x$ and $X_{e_x} \in \mathsf{T}_{e_x}\mathsf{E}$ we have

$$T_{e_x}\hat{\iota}_{\mathsf{E}}(X_{e_x}) \in \mathsf{T}_{\hat{\iota}_{\mathsf{E}}(e_x)}(\mathbb{R}^N \times \mathbb{R}^N) \simeq \mathbb{R}^N \oplus \mathbb{R}^N,$$

and we define K so that

$$\hat{\iota}_{\mathsf{E}} \circ K(X_{e_x}) = \mathrm{pr}_2 \circ T_{e_x}\hat{\iota}_{\mathsf{E}}(X_{e_x});$$

this uniquely defines K by injectivity of $\hat{\iota}_{\mathsf{E}}$, and amounts to using on E the connection induced on image($\hat{\iota}_{\mathsf{E}}$) by the trivial connection on $\mathbb{R}^N \times \mathbb{R}^N$. In particular, this means that we think of $\hat{\iota}_{\mathsf{E}} \circ K(X_{e_x})$ as being an element of the fibre of the trivial bundle $\mathbb{R}^N \times \mathbb{R}^N$ at $\iota_{\mathsf{M}}(x)$.

If $v_x \in \mathsf{TM}$, if $e, e' \in \mathsf{E}$, and if $X \in \mathsf{T}_e\mathsf{E}$ and $X' \in \mathsf{T}_{e'}\mathsf{E}$ satisfy $X, X' \in T\pi^{-1}(v_x)$, then note that

$$\begin{aligned} T_e\pi(X) = T_{e'}\pi(X') &\implies T_e(\iota_{\mathsf{M}} \circ \pi)(X) = T_{e'}(\iota_{\mathsf{M}} \circ \pi)(X') \\ &\implies T_e(\mathrm{pr}_2 \circ \hat{\iota}_{\mathsf{E}})(X) = T_{e'}(\mathrm{pr}_2 \circ \hat{\iota}_{\mathsf{E}})(X') \\ &\implies T_{\iota_{\mathsf{M}}(x)}\,\mathrm{pr}_2 \circ T_e\hat{\iota}_{\mathsf{E}}(X) = T_{\iota_{\mathsf{M}}(x)}\,\mathrm{pr}_2 \circ T_{e'}\hat{\iota}_{\mathsf{E}}(X'). \end{aligned}$$

Thus we can write

$$T_e\hat{\iota}_{\mathsf{E}}(X) = (\boldsymbol{x}, \boldsymbol{e}, \boldsymbol{u}, \boldsymbol{v}), \quad T_{e'}\hat{\iota}_{\mathsf{E}}(X) = (\boldsymbol{x}, \boldsymbol{e}', \boldsymbol{u}, \boldsymbol{v}')$$

for suitable $\boldsymbol{x}, \boldsymbol{u}, \boldsymbol{e}, \boldsymbol{e}', \boldsymbol{v}, \boldsymbol{v}' \in \mathbb{R}^N$. Therefore,

$$\hat{\iota}_{\mathsf{E}} \circ K(X) = (\boldsymbol{x}, \boldsymbol{v}), \quad \hat{\iota}_{\mathsf{E}} \circ K(X') = (\boldsymbol{x}, \boldsymbol{v}'), \quad \hat{\iota}_{\mathsf{E}} \circ K(X + X') = (\boldsymbol{x}, \boldsymbol{v} + \boldsymbol{v}'),$$

from which we immediately conclude that, for addition in the vector bundle $T\pi\colon \mathsf{TE} \to \mathsf{TM}$, we have

$$\hat{\iota}_{\mathsf{E}} \circ K(X + X') = \hat{\iota}_{\mathsf{E}} \circ K(X) + \hat{\iota}_{\mathsf{E}} \circ K(X'),$$

showing that the diagram on the left in (2.8) makes K a vector bundle mapping.

On the other hand, if $e_x \in \mathsf{E}$ and if $X, X' \in \mathsf{T}_{e_x}\mathsf{E}$, then we have, using vector bundle addition in $\pi_{\mathsf{TE}}: \mathsf{TE} \to \mathsf{E}$,

$$\begin{aligned}\hat{\iota}_{\mathsf{E}} \circ K(X + X') &= \mathrm{pr}_2 \circ T_{e_x}\hat{\iota}_{\mathsf{E}}(X + X') \\ &= \mathrm{pr}_2 \circ T_{e_x}\hat{\iota}_{\mathsf{E}}(X) + \mathrm{pr}_2 \circ T_{e_x}\hat{\iota}_{\mathsf{E}}(X') \\ &= \hat{\iota}_{\mathsf{E}} \circ K(X) + \hat{\iota}_{\mathsf{E}} \circ K(X'),\end{aligned}$$

giving that the diagram on the right in (2.8) makes K a vector bundle mapping. Since K is real analytic, this defines a real analytic linear connection ∇^0 on E as in [7, Sect. 11.11].

The existence of $\mathbb{G}_0$, $\mathbb{G}$, and ∇ are straightforward. Indeed, we let $\mathbb{G}_{\mathbb{R}^N}$ be the Euclidean metric on $\mathbb{R}^N$, and define $\mathbb{G}_0$ and $\mathbb{G}$ by

$$\mathbb{G}_0(e_x, e'_x) = \mathbb{G}_{\mathbb{R}^N}(\hat{\iota}_{\mathsf{E}}(e_x), \hat{\iota}_{\mathsf{E}}(e'_x))$$

and

$$\mathbb{G}(v_x, v'_x) = \mathbb{G}_{\mathbb{R}^N}(T_x\iota_{\mathsf{M}}(v_x), T_x\iota_{\mathsf{M}}(v'_x)).$$

The affine connection ∇ can be taken to be the Levi–Civita connection of $\mathbb{G}$. □

The existence of a real analytic linear connection in a real analytic vector bundle is asserted at the bottom of page 302 in [9], and we fill in the blanks in the preceding proof.

Now let us provide a fundamental relationship between the geometric fibre norms of Sect. 2.2 and norms constructed in local coordinate charts.

Lemma 2.5 (A Fundamental Estimate for Fibre Norms). *Let $\mathcal{U} \subseteq \mathbb{R}^n$ be open, denote $\mathbb{R}^k_{\mathcal{U}} = \mathcal{U} \times \mathbb{R}^k$, let $K \subseteq \mathcal{U}$ be compact, and consider the trivial vector bundle $\mathrm{pr}_1: \mathbb{R}^k_{\mathcal{U}} \to \mathcal{U}$. Let $\mathbb{G}$ be a Riemannian metric on $\mathcal{U}$, let $\mathbb{G}_0$ be a vector bundle metric on $\mathbb{R}^k_{\mathcal{U}}$, let ∇ be an affine connection on $\mathcal{U}$, and let ∇^0 be a vector bundle connection on $\mathbb{R}^k_{\mathcal{U}}$, with all of these being real analytic. Then there exist $C, \sigma \in \mathbb{R}_{>0}$ such that*

$$\frac{\sigma^m}{C}\|j_m\boldsymbol{\xi}(\boldsymbol{x})\|_{\mathbb{G}_m} \le \sup\left\{\frac{1}{I!}|\boldsymbol{D}^I\xi^a(\boldsymbol{x})| \;\middle|\; |I| \le m,\ a \in \{1,\dots,k\}\right\} \le \frac{C}{\sigma^m}\|j_m\boldsymbol{\xi}(\boldsymbol{x})\|_{\mathbb{G}_m}$$

for every $\boldsymbol{\xi} \in \Gamma^\infty(\mathbb{R}^k_{\mathcal{U}})$, $\boldsymbol{x} \in K$, and $m \in \mathbb{Z}_{\ge 0}$.

Proof: As the proof is quite complicated, let us first provide an outline. First, fixing $\boldsymbol{x} \in \mathcal{U}$, we use Lemma 2.2 to establish a relationship between jets, decomposed according to Lemma 2.1, and ordinary derivatives in a local trivialisation associated with the connections ∇ and ∇^0. This is done in Sublemma 5. This estimate itself involves estimates associated with local vector bundle isomorphisms, which are stated as Sublemma 4. The sublemma, in turn relies on estimates associated with composition and multiplication of real analytic functions; these are given as Sublemmata 1, 2, and 3. After establishing an estimate of the desired form at a fixed $\boldsymbol{x} \in \mathcal{U}$, we note that, in a neighbourhood of $\boldsymbol{x}$, a local section associated with a local section $\boldsymbol{\xi}$ by the vector bundle isomorphism of Lemma 2.2 can be thought of as a parameterised family of local sections in a neighbourhood of $\boldsymbol{x}$. By this device we are able to "smear" the pointwise estimate of Sublemma 5 to a neighbourhood of x; this is done in Sublemma 8, which itself is stated after introducing some straightforward, but elementary, constructions based on Lemma 2.2. After this, the final estimates needed for the proof of the lemma are easily made.

We shall thus first consider estimates associated with local vector bundle maps. First we consider an estimate arising from multiplication.

Sublemma 1 *If $\mathcal{U} \subseteq \mathbb{R}^n$ is open, if $f \in C^\omega(\mathcal{U})$, and if $K \subseteq \mathcal{U}$ is compact, then there exist $C, \sigma \in \mathbb{R}_{>0}$ such that*

$$\sup\left\{\frac{1}{I!}\boldsymbol{D}^I(fg)(\boldsymbol{x}) \;\middle|\; |I| \leq m\right\} \leq C\sigma^{-m} \sup\left\{\frac{1}{I!}\boldsymbol{D}^I g(\boldsymbol{x}) \;\middle|\; |I| \leq m\right\}$$

for every $g \in C^\infty(\mathcal{U})$, $\boldsymbol{x} \in K$, and $m \in \mathbb{Z}_{\geq 0}$.

Proof: For multi-indices $I, J \in \mathbb{Z}^n_{\geq 0}$, let us write $J \leq I$ if $I - J \in \mathbb{Z}^n_{\geq 0}$. For $I \in \mathbb{Z}^n_{\geq 0}$ we have

$$\frac{1}{I!}\boldsymbol{D}^I(fg)(\boldsymbol{x}) = \sum_{J \leq I} \frac{\boldsymbol{D}^J g(\boldsymbol{x})}{J!} \frac{\boldsymbol{D}^{I-J} f(\boldsymbol{x})}{(I-J)!},$$

by the Leibniz Rule. By [8, Lemma 2.1.3], the number of multi-indices in n variables of order at most $|I|$ is $\frac{(n+|I|)!}{n!|I|!}$. Note that, by the binomial theorem,

$$(a_1 + a_2)^{n+|I|} = \sum_{j=0}^{n+|I|} \frac{(n+|I|)!}{(n+|I|-j)!j!} a_1^j a_2^{n+|I|-j}.$$

Evaluating at $a_1 = a_2 = 1$ and considering the summand corresponding to $j = |I|$, this gives

$$\frac{(n+|I|)!}{n!|I|!} \leq 2^{n+|I|}.$$

Using this inequality we derive

$$\begin{aligned}
&\frac{1}{I!}|\boldsymbol{D}^I(fg)(\boldsymbol{x})| \\
&\leq \sum_{|J| \leq |I|} \sup\left\{\frac{|\boldsymbol{D}^J f(\boldsymbol{x})|}{J!} \;\middle|\; |J| \leq |I|\right\} \sup\left\{\frac{|\boldsymbol{D}^J g(\boldsymbol{x})|}{J!} \;\middle|\; |J| \leq |I|\right\} \\
&\leq \frac{(n+|I|)!}{n!|I|!} \sup\left\{\frac{|\boldsymbol{D}^J f(\boldsymbol{x})|}{J!} \;\middle|\; |J| \leq |I|\right\} \sup\left\{\frac{|\boldsymbol{D}^J g(\boldsymbol{x})|}{J!} \;\middle|\; |J| \leq |I|\right\} \\
&\leq 2^{n+|I|} \sup\left\{\frac{|\boldsymbol{D}^J f(\boldsymbol{x})|}{J!} \;\middle|\; |J| \leq |I|\right\} \sup\left\{\frac{|\boldsymbol{D}^J g(\boldsymbol{x})|}{J!} \;\middle|\; |J| \leq |I|\right\}.
\end{aligned}$$

By [8, Proposition 2.2.10], there exist $B, r \in \mathbb{R}_{>0}$ such that

$$\frac{1}{J!}|\boldsymbol{D}^J f(\boldsymbol{x})| \leq B r^{-|J|}, \qquad J \in \mathbb{Z}^n_{\geq 0},\ \boldsymbol{x} \in K.$$

We can suppose, without loss of generality, that $r < 1$ so that we have

$$\frac{1}{I!}|\boldsymbol{D}^I(fg)(\boldsymbol{x})| \leq 2^n B\Big(\frac{2}{r}\Big)^{|I|} \sup\Big\{\frac{|\boldsymbol{D}^J g(\boldsymbol{x})|}{J!} \;\Big|\; |J| \leq |I|\Big\}, \qquad \boldsymbol{x} \in K.$$

We conclude, therefore, that if $|I| \leq m$ we have

$$\frac{1}{I!}|\boldsymbol{D}^I(fg)(\boldsymbol{x})| \leq 2^n B\Big(\frac{2}{r}\Big)^{m} \sup\Big\{\frac{|\boldsymbol{D}^J g(\boldsymbol{x})|}{J!} \;\Big|\; |J| \leq m\Big\}, \qquad \boldsymbol{x} \in K,$$

which is the result upon taking $C = 2^n B$ and $\sigma = \frac{2}{r}$. ▽

Next we give an estimate for derivatives of compositions of mappings, one of which is real analytic. Thus we have a real analytic mapping $\boldsymbol{\Phi}\colon \mathcal{U} \to \mathcal{V}$ between open sets $\mathcal{U} \subseteq \mathbb{R}^n$ and $\mathcal{V} \subseteq \mathbb{R}^k$ and $f \in \mathrm{C}^\infty(\mathcal{V})$. By the higher-order Chain Rule, e.g., [2], we can write

$$\boldsymbol{D}^I(f \circ \boldsymbol{\Phi})(\boldsymbol{x}) = \sum_{\substack{H \in \mathbb{Z}^m_{\geq 0} \\ |H| \leq |I|}} A_{I,H}(\boldsymbol{x}) \boldsymbol{D}^H f(\boldsymbol{\Phi}(\boldsymbol{x}))$$

for $\boldsymbol{x} \in \mathcal{U}$ and for some real analytic functions $A_{I,H} \in \mathrm{C}^\omega(\mathcal{U})$. The proof of the next sublemma gives estimates for the $A_{I,H}$'s and is based on computations of Thilliez [13] in the proof of his Proposition 2.5.

Sublemma 2 *Let $\mathcal{U} \subseteq \mathbb{R}^n$ and $\mathcal{V} \subseteq \mathbb{R}^k$ be open, let $\boldsymbol{\Phi} \in \mathrm{C}^\omega(\mathcal{U}; \mathcal{V})$, and let $K \subseteq \mathcal{U}$ be compact. Then there exist $C, \sigma \in \mathbb{R}_{>0}$ such that*

$$|\boldsymbol{D}^J A_{I,H}(\boldsymbol{x})| \leq C\sigma^{-(|I|+|J|)}(|I| + |J| - |H|)!$$

for every $\boldsymbol{x} \in K$, $I, J \in \mathbb{Z}^n_{\geq 0}$, and $H \in \mathbb{Z}^k_{\geq 0}$.

Proof: First we claim that, for $j_1, \dots, j_r \in \{1, \dots, n\}$,

$$\frac{\partial^r (f \circ \boldsymbol{\Phi})}{\partial x^{j_1} \cdots \partial x^{j_r}}(\boldsymbol{x}) = \sum_{s=1}^{r} \sum_{a_1,\dots,a_s=1}^{k} B^{a_1\cdots a_s}_{j_1\cdots j_r}(\boldsymbol{x}) \frac{\partial^s f}{\partial y^{a_1} \cdots \partial y^{a_s}}(\boldsymbol{\Phi}(\boldsymbol{x})),$$

where the real analytic functions $B^{a_1\cdots a_s}_{j_1\cdots j_r}$, $a_1, \dots, a_s \in \{1, \dots, k\}$, $j_1, \dots, j_r \in \{1, \dots, n\}$, $r, s \in \mathbb{Z}_{>0}$, $s \leq r$, are defined by the following recursion, starting with $B^a_j = \frac{\partial \Phi^a}{\partial x^j}$:

1. $B^a_{j_1\cdots j_r} = \dfrac{\partial B^a_{j_2\cdots j_r}}{\partial x^{j_1}}$;
2. $B^{a_1\cdots a_s}_{j_1\cdots j_r} = \dfrac{\partial B^{a_1\cdots a_s}_{j_2\cdots j_r}}{\partial x^{j_1}} + \dfrac{\partial \Phi^{a_1}}{\partial x^{j_1}} B^{a_2\cdots a_s}_{j_2\cdots j_r}$, $r \geq 2$, $s \in \{2, \dots, r-1\}$;
3. $B^{a_1\cdots a_r}_{j_1\cdots j_r} = \dfrac{\partial \Phi^{a_1}}{\partial x^{j_1}} B^{a_2\cdots a_r}_{j_2\cdots j_r}$.

This claim we prove by induction on r. It is clear for $r = 1$, so suppose the assertion true up to $r - 1$. By the induction hypothesis we have

$$\frac{\partial^{r-1}(f \circ \boldsymbol{\Phi})}{\partial x^{j_2} \cdots \partial x^{j_r}}(\boldsymbol{x}) = \sum_{s=1}^{r-1} \sum_{a_1,\dots,a_s=1}^{k} B^{a_1 \cdots a_s}_{j_2 \cdots j_r}(\boldsymbol{x}) \frac{\partial^s f}{\partial y^{a_1} \cdots \partial y^{a_s}}(\boldsymbol{\Phi}(\boldsymbol{x})).$$

We then compute

$$\begin{aligned}
&\frac{\partial}{\partial x^{j_1}} \frac{\partial^{r-1}(f \circ \boldsymbol{\Phi})}{\partial x^{j_2} \cdots \partial x^{j_r}}(\boldsymbol{x}) \\
&= \sum_{s=1}^{r-1} \sum_{a_1,\dots,a_s=1}^{k} \Big(\frac{\partial B^{a_1,\dots,a_s}_{j_2 \cdots j_r}}{\partial x^{j_1}}(\boldsymbol{x}) \frac{\partial^s f}{\partial y^{a_1} \cdots \partial y^{a_s}}(\boldsymbol{\Phi}(\boldsymbol{x})) \\
&\quad + \sum_{b=1}^{k} B^{a_1 \cdots a_s}_{j_2 \cdots j_r}(\boldsymbol{x}) \frac{\partial \Phi^b}{\partial x^{j_1}}(\boldsymbol{x}) \frac{\partial^{s+1} f}{\partial y^b \partial y^{a_1} \cdots \partial y^{a_s}}(\boldsymbol{\Phi}(\boldsymbol{x})) \Big) \\
&= \sum_{s=1}^{r-1} \sum_{a_1,\dots,a_s=1}^{k} \frac{\partial B^{a_1 \cdots a_s}_{j_2 \cdots j_r}}{\partial x^{j_1}}(\boldsymbol{x}) \frac{\partial^s f}{\partial y^{a_1} \cdots \partial y^{a_s}}(\boldsymbol{\Phi}(\boldsymbol{x})) \\
&\quad + \sum_{s=2}^{r} \sum_{a_1,\dots,a_s=1}^{k} B^{a_2 \cdots a_s}_{j_2 \cdots j_r}(\boldsymbol{x}) \frac{\partial \Phi^{a_1}}{\partial x^{j_1}}(\boldsymbol{x}) \frac{\partial^s f}{\partial y^{a_1} \cdots \partial y^{a_s}}(\boldsymbol{\Phi}(\boldsymbol{x})) \\
&= \sum_{a=1}^{k} \frac{\partial B^{a}_{j_2 \cdots j_r}}{\partial x^{j_1}}(\boldsymbol{x}) \frac{\partial f}{\partial y^a}(\boldsymbol{\Phi}(\boldsymbol{x})) \\
&\quad + \sum_{s=2}^{r-1} \sum_{a_1,\dots,a_s=1}^{k} \Big(\frac{\partial B^{a_1 \cdots a_s}_{j_2 \cdots j_r}}{\partial x^{j_1}}(\boldsymbol{x}) + \frac{\partial \Phi^{a_1}}{\partial x^{j_1}}(\boldsymbol{x}) B^{a_2 \cdots a_s}_{j_2 \cdots j_r}(\boldsymbol{x}) \Big) \frac{\partial^s f}{\partial y^{a_1} \cdots \partial y^{a_s}}(\boldsymbol{\Phi}(\boldsymbol{x})) \\
&\quad + \sum_{a_1,\dots,a_r=1}^{k} \frac{\partial \Phi^{a_1}}{\partial x^{j_1}}(\boldsymbol{x}) B^{a_2 \cdots a_r}_{j_2 \cdots j_r}(\boldsymbol{x}) \frac{\partial^s f}{\partial y^{a_1} \cdots \partial y^{a_r}}(\boldsymbol{\Phi}(\boldsymbol{x})),
\end{aligned}$$

from which our claim follows.

Next we claim that there exist $A, \rho, \alpha, \beta \in \mathbb{R}_{>0}$ such that

$$|\boldsymbol{D}^J B^{a_1 \cdots a_s}_{j_1 \cdots j_r}(\boldsymbol{x})| \leq (A\alpha)^r \Big(\frac{\beta}{\rho}\Big)^{r+|J|-s} (r + |J| - s)!$$

for every $\boldsymbol{x} \in K$, $J \in \mathbb{Z}^n_{\geq 0}$, $a_1, \dots, a_s \in \{1, \dots, k\}$, $j_1, \dots, j_k \in \{1, \dots, n\}$, $r, s \in \mathbb{Z}_{>0}$, $s \leq r$. This we prove by induction on r once again. First let $\beta \in \mathbb{R}_{>0}$ be sufficiently large that

$$\sum_{I \in \mathbb{Z}^n_{\geq 0}} \beta^{-|I|} < \infty,$$

and denote this value of this sum by S. Then let $\alpha = 2S$. By [8, Proposition 2.2.10] there exist $A, \rho \in \mathbb{R}_{>0}$ such that

$$|\boldsymbol{D}^J \boldsymbol{D}^j \Phi^a(\boldsymbol{x})| \leq AJ!\rho^{-|J|}$$

for every $\boldsymbol{x} \in K$, $J \in \mathbb{Z}^n_{\geq 0}$, $j \in \{1, \ldots, n\}$, and $a \in \{1, \ldots, k\}$. This gives the claim for $r = 1$. So suppose the claim true up to $r - 1$. Then, for any $a_1, \ldots, a_s \in \{1, \ldots, k\}$ and $j_1, \ldots, j_r \in \{1, \ldots, n\}$, $s \leq r$, $B^{a_1 \cdots a_s}_{j_1 \cdots j_r}$ has one of the three forms listed above in the recurrent definition. These three forms are themselves sums of terms of the form

$$\underbrace{\frac{\partial B^{a_1 \cdots a_s}_{j_2 \cdots j_r}}{\partial x^{j_1}}}_{P}, \qquad \underbrace{\frac{\partial \Phi^{a_1}}{\partial x^{j_1}} B^{a_2 \cdots a_s}_{j_2 \cdots j_r}}_{Q}.$$

Let us, therefore, estimate derivatives of these terms, abbreviated by P and Q as above.

We directly have, by the induction hypothesis,

$$\begin{aligned} |\boldsymbol{D}^J P(\boldsymbol{x})| &\leq (A\alpha)^r \Big(\frac{\beta}{\rho}\Big)^{r+|J|-s} (r + |J| - s)! \\ &\leq A^r \alpha^{r-1} S \Big(\frac{\beta}{\rho}\Big)^{r+|J|-s} (r + |J| - s)!, \end{aligned}$$

noting that $\alpha = 2S$. By the Leibniz Rule we have

$$\boldsymbol{D}^J Q(\boldsymbol{x}) = \sum_{J_1 + J_2 = J} \frac{J!}{J_1! J_2!} \boldsymbol{D}^{J_1} \boldsymbol{D}^{j_1} \Phi^{a_1}(\boldsymbol{x}) \boldsymbol{D}^{J_2} B^{a_2 \cdots a_s}_{j_2 \cdots j_r}(\boldsymbol{x}).$$

By the induction hypothesis we have

$$|\boldsymbol{D}^{J_2} B^{a_2 \cdots a_s}_{j_2 \cdots j_r}(\boldsymbol{x})| \leq (A\alpha)^{r-1} \Big(\frac{\beta}{\rho}\Big)^{r+|J_2|-s} (r + |J_2| - s)!$$

for every $\boldsymbol{x} \in K$ and $J_2 \in \mathbb{Z}_{\geq 0}$. Therefore,

$$|\boldsymbol{D}^J Q(\boldsymbol{x})| \leq \sum_{J_1 + J_2 = J} \frac{J!}{J_2!} A (A\alpha)^{r-1} \Big(\frac{\beta}{\rho}\Big)^{r+|J|-s} \beta^{-|J_1|} (r + |J_2| - s)!$$

for every $\boldsymbol{x} \in K$ and $J \in \mathbb{Z}^n_{\geq 0}$. Now note that, for any $a, b, c \in \mathbb{Z}_{>0}$ with $b < c$, we have

$$\frac{(a + b)!}{b!} = (1 + b) \cdots (a + b) < (1 + c) \cdots (a + c) = \frac{(a + c)!}{c!}.$$

Thus, if $L, J \in \mathbb{Z}^n_{\geq 0}$ satisfy $L < J$ (meaning that $J - L \in \mathbb{Z}^n_{\geq 0}$), then we have

$$l_k \leq j_k \quad \implies \quad \frac{(a + l_k)!}{l_k!} \leq \frac{(a + j_k)!}{j_k!} \quad \implies \quad \frac{j_k!}{l_k!} \leq \frac{(a + j_k)!}{(a + l_k)!}$$

for every $a \in \mathbb{Z}_{>0}$ and $k \in \{1, \dots, n\}$. Therefore,

$$\frac{(j_1 + \cdots + j_{n-1} + j_n)!}{(j_1 + \cdots + j_{n-1} + l_n)!} \geq \frac{j_n!}{l_n!}$$

and

$$\begin{aligned}\frac{(j_1 + \cdots + j_{n-2} + j_{n-1} + j_n)!}{(j_1 + \cdots + j_{n-2} + l_{n-1} + l_n)!} \\ = \frac{(j_1 + \cdots + j_{n-1} + j_n)!}{(j_1 + \cdots + j_{n-1} + l_n)!} \frac{(j_1 + \cdots + j_{n-2} + j_{n-1} + l_n)!}{(j_1 + \cdots + j_{n-2} + l_{n-1} + l_n)!} \geq \frac{j_{n-1}!}{l_{n-1}!} \frac{j_n!}{l_n!}.\end{aligned}$$

Continuing in this way, we get

$$\frac{J!}{L!} \leq \frac{|J|!}{|L|!}.$$

We also have

$$\frac{(r + |J_2| - s)!}{|J_2|!} \leq \frac{(r + |J| - s)!}{|J|!}.$$

Thus we have

$$\begin{aligned}|\boldsymbol{D}^J Q(\boldsymbol{x})| &\leq \sum_{J_1+J_2=J} \frac{J!}{J_2!} A(A\alpha)^{r-1} \Big(\frac{\beta}{\rho}\Big)^{r+|J|-s} \beta^{-|J_1|} (r + |J_2| - s)! \\ &\leq A(A\alpha)^{r-1} \Big(\frac{\beta}{\rho}\Big)^{r+|J|-s} (r + |J| - s)! \sum_{J_1+J_2=J} \beta^{-|J_1|} \\ &\leq AS\,(A\alpha)^{r-1} \Big(\frac{\beta}{\rho}\Big)^{r+|J|-s} (r + |J| - s)!.\end{aligned}$$

Combining the estimates for P and Q to give an estimate for their sum, and recalling that $\alpha = 2S$, gives our claim that there exist $A, \rho, \alpha, \beta \in \mathbb{R}_{>0}$ such that

$$|\boldsymbol{D}^J B^{a_1 \cdots a_s}_{j_1 \cdots j_r}(\boldsymbol{x})| \leq (A\alpha)^r \Big(\frac{\beta}{\rho}\Big)^{r+|J|-s} (r + |J| - s)!$$

for every $\boldsymbol{x} \in K$, $J \in \mathbb{Z}^n_{\geq 0}$, $a_1, \dots, a_s \in \{1, \dots, k\}$, and $j_1, \dots, j_r \in \{1, \dots, n\}$, $r, s \in \mathbb{Z}_{>0}$, $s \leq r$.

To conclude the proof of the lemma, note that given an index $\boldsymbol{j} = (j_1, \dots, j_r) \in \{1, \dots, n\}^r$ we define a multi-index $I(\boldsymbol{j}) = (i_1, \dots, i_n) \in \mathbb{Z}^n_{\geq 0}$ by asking that i_l be the number of times that l appears in the list $\boldsymbol{j}$. Similarly an index $\boldsymbol{a} = (a_1, \dots, a_s) \in \{1, \dots, k\}^s$ gives rise to a multi-index $H(\boldsymbol{a}) \in \mathbb{Z}^k_{\geq 0}$. Moreover, by construction we have

$$B^{a_1 \cdots a_s}_{j_1 \cdots j_r} = A_{I(\boldsymbol{j}), H(\boldsymbol{a})}.$$

Let $C = 1$ and $\sigma^{-1} = \max\{A\alpha, \frac{\beta}{\rho}\}$ and suppose, without loss of generality, that $\sigma \leq 1$. Then

$$(A\alpha)^{|I|} \leq \sigma^{-(|I|+|J|)}, \quad \Big(\frac{\beta}{\rho}\Big)^{r+|J|-s} \leq \sigma^{-(|I|+|J|)}$$

for every $I, J \in \mathbb{Z}^n_{\geq 0}$. Thus we have

$$|\boldsymbol{D}^J A_{I,H}(\boldsymbol{x})| \leq C\sigma^{-(|I|+|J|)}(|I| + |J| - |H|)!$$

as claimed. ▽

Next we consider estimates for derivatives arising from composition.

Sublemma 3 *Let $\mathcal{U} \subseteq \mathbb{R}^n$ and $\mathcal{V} \subseteq \mathbb{R}^k$ be open, let $\boldsymbol{\Phi} \in \mathrm{C}^\omega(\mathcal{U}; \mathcal{V})$, and let $K \subseteq \mathcal{U}$ be compact. Then there exist $C, \sigma \in \mathbb{R}_{>0}$ such that*

$$\sup\left\{\frac{1}{I!}|\boldsymbol{D}^I(f \circ \boldsymbol{\Phi})(\boldsymbol{x})| \;\Big|\; |I| \leq m\right\} \leq C\sigma^{-m} \sup\left\{\frac{1}{I!}|\boldsymbol{D}^H f(\boldsymbol{\Phi}(\boldsymbol{x}))| \;\Big|\; |H| \leq m\right\}$$

for every $f \in \mathrm{C}^\infty(\mathcal{V})$, $\boldsymbol{x} \in K$, and $m \in \mathbb{Z}_{\geq 0}$.

Proof: As we denoted preceding the statement of Sublemma 2 above, let us write

$$\boldsymbol{D}^I(f \circ \boldsymbol{\Phi})(\boldsymbol{x}) = \sum_{\substack{H \in \mathbb{Z}^m_{\geq 0} \\ |H| \leq |I|}} A_{I,H}(\boldsymbol{x}) \boldsymbol{D}^H f(\boldsymbol{\Phi}(\boldsymbol{x}))$$

for $\boldsymbol{x} \in \mathcal{U}$ and for some real analytic functions $A_{I,H} \in \mathrm{C}^\omega(\mathcal{U})$. By Sublemma 2, let $A, r \in \mathbb{R}_{>0}$ be such that

$$|\boldsymbol{D}^J A_{I,H}(\boldsymbol{x})| \leq Ar^{-(|I|+|J|)}(|I| + |J| - |H|)!$$

for $\boldsymbol{x} \in K$. By the multinomial theorem [8, Theorem 1.3.1] we can write

$$(a_1 + \cdots + a_n)^{|I|} = \sum_{|J|=|I|} \frac{|J|!}{J!} \boldsymbol{a}^J$$

for every $I \in \mathbb{Z}^n_{\geq 0}$. Setting $a_1 = \cdots = a_n = 1$ gives $\frac{|I|!}{I!} \leq n^{|I|}$ for every $I \in \mathbb{Z}^n_{\geq 0}$. As in the proof of Sublemma 1 we have that the number of multi-indices of length k and degree at most $|I|$ is bounded above by $2^{k+|I|}$. Also, by a similar binomial theorem argument, if $|H| \leq |I|$, then we have

$$\frac{(|I| - |H|)!|H|!}{|I|!} \leq 2^{|I|}.$$

Putting this together yields

$$\begin{aligned}
\frac{1}{I!}|\boldsymbol{D}^I(f \circ \boldsymbol{\Phi})(\boldsymbol{x})| &\leq An^{|I|}r^{-|I|} \sum_{|H| \leq |I|} \frac{(|I| - |H|)!|H|!}{|I|!} \frac{1}{H!}|\boldsymbol{D}^H f(\boldsymbol{\Phi}(\boldsymbol{x}))| \\
&\leq An^{|I|}2^{k+|I|}2^{|I|}r^{-|I|} \sup\left\{\frac{1}{H!}|\boldsymbol{D}^H f(\boldsymbol{\Phi}(\boldsymbol{x}))| \;\Big|\; |H| \leq |I|\right\} \\
&= 2^k A(4nr^{-1})^{|I|} \sup\left\{\frac{1}{H!}|\boldsymbol{D}^H f(\boldsymbol{\Phi}(\boldsymbol{x}))| \;\Big|\; |H| \leq |I|\right\}
\end{aligned}$$

whenever $x \in K$. Let us denote $C = 2^k A$ and $\sigma^{-1} = 4nr^{-1}$ and take r so that $4nr^{-1} \geq 1$, without loss of generality. We then have

$$\sup\left\{\frac{1}{I!}|\boldsymbol{D}^I(f \circ \boldsymbol{\Phi})(\boldsymbol{x})| \;\middle|\; |I| \leq m\right\} \leq C\sigma^{-1} \sup\left\{\frac{1}{H!}|\boldsymbol{D}^H f(\boldsymbol{\Phi}(\boldsymbol{x}))| \;\middle|\; |H| \leq m\right\}$$

for every $f \in \mathrm{C}^\infty(\mathcal{U}_2)$, $\boldsymbol{x} \in K$, and $m \in \mathbb{Z}_{\geq 0}$, as claimed. ▽

Now we can state the following estimate for vector bundle mappings which is essential for our proof.

Sublemma 4 *Let $\mathcal{U} \subseteq \mathbb{R}^n$ and $\mathcal{V} \subseteq \mathbb{R}^k$ be open, let $l \in \mathbb{Z}_{>0}$, and consider the trivial vector bundles $\mathbb{R}^l_{\mathcal{U}}$ and $\mathbb{R}^l_{\mathcal{V}}$. Let $\boldsymbol{\Phi} \in \mathrm{C}^\omega(\mathcal{U}; \mathcal{V})$, let $\boldsymbol{A} \in \mathrm{C}^\omega(\mathcal{U}; \mathsf{GL}(l; \mathbb{R}))$, and let $K \subseteq \mathcal{U}$ be compact. Then there exist $C, \sigma \in \mathbb{R}_{>0}$ such that*

$$\sup\left\{\frac{1}{I!}|\boldsymbol{D}^I(\boldsymbol{A}^{-1} \cdot (\boldsymbol{\xi} \circ \boldsymbol{\Phi}))^b(\boldsymbol{x})| \;\middle|\; |I| \leq m,\ b \in \{1, \dots, l\}\right\}$$
$$\leq C\sigma^{-m} \sup\left\{\frac{1}{H!}|\boldsymbol{D}^H \xi^a(\boldsymbol{\Phi}(\boldsymbol{x}))| \;\middle|\; |H| \leq m,\ a \in \{1, \dots, l\}\right\},$$

for every $\boldsymbol{\xi} \in \Gamma^\infty(\mathbb{R}^l_{\mathcal{V}})$, $\boldsymbol{x} \in K$, and $m \in \mathbb{Z}_{\geq 0}$.

Proof: By Sublemma 3 there exist $C_1, \sigma_1 \in \mathbb{R}_{>0}$ such that

$$\sup\left\{\frac{1}{I!}|\boldsymbol{D}^I(\boldsymbol{\xi} \circ \boldsymbol{\Phi})^a(\boldsymbol{x})| \;\middle|\; |I| \leq m,\ a \in \{1, \dots, l\}\right\}$$
$$\leq C_1\sigma_1^{-m} \sup\left\{\frac{1}{H!}|\boldsymbol{D}^H \xi^a(\boldsymbol{\Phi}(\boldsymbol{x}))| \;\middle|\; |H| \leq m,\ a \in \{1, \dots, l\}\right\}$$

for every $\boldsymbol{\xi} \in \Gamma^\infty(\mathbb{R}^l_{\mathcal{V}})$, $\boldsymbol{x} \in K$, and $m \in \mathbb{Z}_{\geq 0}$.

Now let $\boldsymbol{\eta} \in \Gamma^\infty(\mathbb{R}^l_{\mathcal{U}})$. Let $B^b_a \in \mathrm{C}^\omega(\mathcal{U})$, $a \in \{1, \dots, l\}$, $b \in \{1, \dots, l\}$, be the components of $\boldsymbol{A}^{-1}$. By Sublemma 1, there exist $C_2, \sigma_2 \in \mathbb{R}_{>0}$ such that

$$\sup\left\{\frac{1}{I!}|\boldsymbol{D}^I(B^b_a(\boldsymbol{x})\eta^a(\boldsymbol{x}))| \;\middle|\; |I| \leq m,\ a, b \in \{1, \dots, l\}\right\}$$
$$\leq C_2\sigma_2^{-m} \sup\left\{\frac{1}{I!}|\boldsymbol{D}^I \eta^a(\boldsymbol{x})| \;\middle|\; |I| \leq m,\ a \in \{1, \dots, l\}\right\}$$

for every $\boldsymbol{x} \in K$ and $m \in \mathbb{Z}_{\geq 0}$. (There is no implied sum over "a" in the preceding formula.) Therefore, by the triangle inequality,

$$\sup\left\{\frac{1}{I!}|\boldsymbol{D}^I(\boldsymbol{A}^{-1} \cdot \boldsymbol{\eta})^b(\boldsymbol{x})| \;\middle|\; |I| \leq m,\ b \in \{1, \dots, l\}\right\}$$
$$\leq lC_2\sigma_2^{-m} \sup\left\{\frac{1}{I!}|\boldsymbol{D}^I \eta^a(\boldsymbol{x})| \;\middle|\; |I| \leq m,\ a \in \{1, \dots, l\}\right\}$$

for every $\boldsymbol{x} \in K$ and $m \in \mathbb{Z}_{\geq 0}$.

Combining the estimates from the preceding two paragraphs gives

$$\sup\left\{\frac{1}{I!}|\boldsymbol{D}^I(\boldsymbol{\xi}\circ\boldsymbol{\Phi})^b(\boldsymbol{x})| \;\middle|\; |I|\leq m,\ b\in\{1,\dots,l\}\right\}$$
$$\leq lC_1C_2(\sigma_1\sigma_2)^{-m}\sup\left\{\frac{1}{H!}|\boldsymbol{D}^H\xi^a(\boldsymbol{\Phi}(\boldsymbol{x}))| \;\middle|\; |H|\leq m,\ a\in\{1,\dots,l\}\right\}$$

for every $\boldsymbol{\xi}\in\Gamma^\infty(\mathbb{R}^l_{\mathcal{V}})$, $\boldsymbol{x}\in K$, and $m\in\mathbb{Z}_{\geq 0}$, which is the desired result after taking $C = lC_1C_2$ and $\sigma = \sigma_1\sigma_2$. ▽

Now we establish an estimate resembling that of the required form for a fixed $\boldsymbol{x}\in\mathcal{U}$.

Sublemma 5 *Let $\mathcal{U}\subseteq\mathbb{R}^n$ be open, denote $\mathbb{R}^k_{\mathcal{U}} = \mathcal{U}\times\mathbb{R}^k$, and consider the trivial vector bundle $\mathrm{pr}_1:\mathbb{R}^k_{\mathcal{U}}\to\mathcal{U}$. Let $\mathbb{G}$ be a Riemannian metric on $\mathcal{U}$, let $\mathbb{G}_0$ be a vector bundle metric on $\mathbb{R}^k_{\mathcal{U}}$, let ∇ be an affine connection on $\mathcal{U}$, and let ∇^0 be a vector bundle connection on $\mathbb{R}^k_{\mathcal{U}}$, with all of these being real analytic. For $\boldsymbol{\xi}\in\Gamma^\infty(\mathbb{R}^k_{\mathcal{U}})$ and $\boldsymbol{x}\in\mathcal{U}$, denote by $\hat{\boldsymbol{\xi}}_{\boldsymbol{x}}$ the corresponding section of $\mathcal{N}_{\boldsymbol{x}}\times\mathbb{R}^k$ defined by the isomorphism $\kappa_{\boldsymbol{x}}$ of Lemma 2.2. For $K\subseteq\mathcal{U}$ compact, there exist $C,\sigma\in\mathbb{R}_{>0}$ such that the following inequalities hold for each $\boldsymbol{\xi}\in\Gamma^\infty(\mathbb{R}^k_{\mathcal{U}})$, $\boldsymbol{x}\in K$, and $m\in\mathbb{Z}_{\geq 0}$:*

(i) $\|j_m\boldsymbol{\xi}(\boldsymbol{x})\|_{\overline{\mathbb{G}}_m}\leq C\sigma^{-m}\sup\left\{\frac{1}{I!}|\boldsymbol{D}^I\hat{\xi}^a_{\boldsymbol{x}}(\boldsymbol{0})| \;\middle|\; |I|\leq m,\ a\in\{1,\dots,k\}\right\}$;

(ii) $\left\{\frac{1}{I!}|\boldsymbol{D}^I\hat{\xi}^a_{\boldsymbol{x}}(\boldsymbol{0})| \;\middle|\; |I|\leq m,\ a\in\{1,\dots,k\}\right\}\leq C\sigma^{-m}\|j_m\boldsymbol{\xi}(\boldsymbol{x})\|_{\overline{\mathbb{G}}_m}$.

Proof: By Lemma 2.2 we have

$$P^m_{\nabla,\nabla^0}(\boldsymbol{\xi})(\boldsymbol{x}) = \boldsymbol{D}^m\hat{\boldsymbol{\xi}}_{\boldsymbol{x}}(\boldsymbol{0})$$

for every $m\in\mathbb{Z}_{\geq 0}$. Take $m\in\mathbb{Z}_{\geq 0}$. We have

$$\sum_{r=0}^m\frac{1}{(r!)^2}\|P^r_{\nabla,\nabla^0}(\boldsymbol{\xi})(\boldsymbol{x})\|^2_{\mathbb{G}_r}\leq\sum_{r=0}^m\frac{A'A^r}{(r!)^2}\|\boldsymbol{D}^r\hat{\boldsymbol{\xi}}_{\boldsymbol{x}}(\boldsymbol{0})\|^2,$$

where $A'\in\mathbb{R}_{>0}$ depends on $\mathbb{G}_0$, $A\in\mathbb{R}_{>0}$ depends on $\mathbb{G}$, and where $\|\cdot\|$ denotes the 2-norm, i.e., the square root of the sum of squares of components. We can, moreover, assume without loss of generality that $A\geq 1$ so that we have

$$\sum_{r=0}^m\frac{1}{(r!)^2}\|P^r_{\nabla,\nabla^0}(\boldsymbol{\xi})(\boldsymbol{x})\|^2_{\mathbb{G}_r}\leq A'A^m\sum_{r=0}^m\frac{1}{(r!)^2}\|\boldsymbol{D}^r\hat{\boldsymbol{\xi}}_{\boldsymbol{x}}(\boldsymbol{0})\|^2.$$

By [8, Lemma 2.1.3],

$$\mathrm{card}(\{I\in\mathbb{Z}^n_{\geq 0}\mid |I|\leq m\}) = \frac{(n+m)!}{n!m!}.$$

Note that the 2-norm for $\mathbb{R}^N$ is related to the ∞-norm for $\mathbb{R}^N$ by $\|\boldsymbol{a}\|_2 \leq \sqrt{N}\|\boldsymbol{a}\|_\infty$ so that

$$\sum_{r=0}^{m} \frac{1}{(r!)^2}\|\boldsymbol{D}^r\hat{\boldsymbol{\xi}}_{\boldsymbol{x}}(\boldsymbol{0})\|^2 \leq k\frac{(n+m)!}{n!m!}\Big(\sup\Big\{\frac{1}{r!}|\boldsymbol{D}^I\hat{\xi}^a_{\boldsymbol{x}}(\boldsymbol{0})| \;\Big|\; |I| \leq m,\ a \in \{1,\dots,k\}\Big\}\Big)^2.$$

By the binomial theorem, as in the proof of Sublemma 1,

$$\frac{(n+m)!}{n!m!} \leq 2^{n+m}.$$

Thus

$$\|j_m\boldsymbol{\xi}(\boldsymbol{x})\|_{\overline{\mathbb{G}}_m} \leq \sqrt{kA'2^n}\left(\sqrt{2A}\right)^m \sup\Big\{\frac{1}{I!}|\boldsymbol{D}^I\hat{\xi}^a_{\boldsymbol{x}}(\boldsymbol{0})| \;\Big|\; |I| \leq m,\ a \in \{1,\dots,k\}\Big\} \tag{2.9}$$

for every $m \in \mathbb{Z}_{\geq 0}$.

To establish the other estimate asserted in the sublemma, let $\boldsymbol{x} \in K$ and, using the notation of Lemma 2.2, let $\mathcal{N}_{\boldsymbol{x}}$ be a relatively compact neighbourhood of $\boldsymbol{0} \in \mathbb{R}^n \simeq \mathsf{T}_{\boldsymbol{x}}\mathbb{R}^n$ and $\mathcal{V}_{\boldsymbol{x}} \subseteq \mathcal{U}$ be a relatively compact neighbourhood of $\boldsymbol{x}$ such that $\kappa_{\boldsymbol{x}}\colon \mathcal{N}_{\boldsymbol{x}} \times \mathbb{R}^k \to \mathcal{V}_{\boldsymbol{x}} \times \mathbb{R}^k$ is a real analytic vector bundle isomorphism. Let $\boldsymbol{\xi} \in \Gamma^\infty(\mathbb{R}^k_{\mathcal{V}_{\boldsymbol{x}}})$ and let $\hat{\boldsymbol{\xi}}_{\boldsymbol{x}} \in \Gamma^\infty(\mathbb{R}^k_{\mathcal{N}_{\boldsymbol{x}}})$ be defined by $\hat{\boldsymbol{\xi}}_{\boldsymbol{x}}(\boldsymbol{v}) = \kappa_{\boldsymbol{x}}^{-1} \circ \boldsymbol{\xi}(\exp_{\boldsymbol{x}}(\boldsymbol{v}))$. As in the first part of the estimate, we have

$$\boldsymbol{D}^m\hat{\boldsymbol{\xi}}_{\boldsymbol{x}}(\boldsymbol{0}) = P^m_{\nabla,\nabla^0}(\boldsymbol{\xi})(\boldsymbol{x})$$

for every $m \in \mathbb{Z}_{\geq 0}$. For indices $\boldsymbol{j} = (j_1,\dots,j_m) \in \{1,\dots,n\}^m$ we define $I(\boldsymbol{j}) = (i_1,\dots,i_n) \in \mathbb{Z}^n_{\geq 0}$ by asking that i_j be the number of times that "j" appears in the list $\boldsymbol{j}$. We then have

$$\begin{aligned}\sup\Big\{\frac{1}{I!}|\boldsymbol{D}^I\hat{\xi}^a_{\boldsymbol{x}}(\boldsymbol{0})| \;\Big|\; |I| \leq m,\ a \in \{1,\dots,k\}\Big\}& \\ = \sup\Big\{\frac{1}{I(\boldsymbol{j})!}|(P^r_{\nabla,\nabla^0}(\boldsymbol{\xi})(\boldsymbol{x}))^a_{j_1\cdots j_r}| \;\Big|\;& \\ j_1,\dots,j_r \in \{1,\dots,n\},\ r \in \{0,1,\dots,m\},\ a \in \{1,\dots,m\}\Big\}&.\end{aligned}$$

By an application of the multinomial theorem as in the proof of Sublemma 3, we have $\frac{|I|!}{I!} \leq n^{|I|}$ for every $I \in \mathbb{Z}^n_{\geq 0}$. We then have

$$\frac{1}{I(\boldsymbol{j})!}|(P^r_{\nabla,\nabla^0}(\boldsymbol{\xi})(\boldsymbol{x}))^a_{j_1\cdots j_r}| \leq \frac{n^r}{r!}|(P^r_{\nabla,\nabla^0}(\boldsymbol{\xi})(\boldsymbol{x}))^a_{j_1\cdots j_r}|$$

for every $j_1,\dots,j_r \in \{1,\dots,n\}$ and $a \in \{1,\dots,k\}$. Using the fact that the ∞-norm for $\mathbb{R}^N$ is related to the 2-norm for $\mathbb{R}^N$ by $\|\boldsymbol{a}\|_\infty \leq \|\boldsymbol{a}\|_2$, we have

$$\sup\Big\{\frac{1}{I!}|\boldsymbol{D}^I\hat{\xi}^a_{\boldsymbol{x}}(\boldsymbol{0})| \;\Big|\; |I| \leq m,\ a \in \{1,\dots,k\}\Big\} \leq \left(\sum_{r=0}^{m}\left(\frac{n^r}{r!}\right)^2 B'B^r\|P^r_{\nabla,\nabla^0}(\boldsymbol{\xi})(\boldsymbol{x})\|^2_{\mathbb{G}_r}\right)^{1/2},$$

where $B' \in \mathbb{R}_{>0}$ depends on $\mathbb{G}_0$ and $B \in \mathbb{R}_{>0}$ depends on $\mathbb{G}$. We may, without loss of generality, suppose that $B \geq 1$ so that we have

$$\sup\left\{\frac{1}{I!}|\boldsymbol{D}^I\hat{\xi}_x^a(\boldsymbol{0})| \;\middle|\; |I| \leq m,\ a \in \{1,\dots,k\}\right\} \leq \sqrt{B'}\left(n\sqrt{B}\right)^m \|j_m\boldsymbol{\xi}(\boldsymbol{x})\|_{\overline{\mathbb{G}}_m}$$

for every $m \in \mathbb{Z}_{\geq 0}$.

The sublemma follows by taking

$$C = \max\left\{\sqrt{kA'2^n},\ \sqrt{B'}\right\}, \quad \sigma^{-1} = \max\left\{\sqrt{2}A, n\sqrt{B}\right\}. \qquad \nabla$$

Next we extend the pointwise estimate of the preceding sublemma to a local estimate. We first introduce some notation in the general setting of Lemma 2.2 that will be useful later. Thus, for the next few paragraphs, we work with a real analytic vector bundle $\pi\colon \mathsf{E} \to \mathsf{M}$ with ∇ an affine connection on M and ∇^0 a linear connection in E. We fix $x \in \mathsf{M}$. We let $\mathcal{N}_x \subseteq \mathsf{T}_x\mathsf{M}$ and $\mathcal{V}_x \subseteq \mathsf{M}$ be neighbourhoods of 0_x and x, respectively, such that $\exp_x\colon \mathcal{N}_x \to \mathcal{V}_x$ is a diffeomorphism. For $y \in \mathcal{V}_x$ we then define

$$\begin{aligned} I'_{xy}\colon \mathcal{N}'_{xy} \times \mathsf{E}_x &\to \mathsf{E}|\mathcal{V}'_{xy} \\ (v, e_x) &\mapsto \tau_{x,\exp_x(v+\exp_x^{-1}(y))}(e_x) \end{aligned}$$

for neighbourhoods $\mathcal{N}'_{xy} \subseteq \mathsf{T}_x\mathsf{M}$ of $0_x \in \mathsf{T}_x\mathsf{M}$ and $\mathcal{V}'_{xy} \subseteq \mathsf{M}$ of y. We note that I'_{xy} is a real analytic vector bundle isomorphism over the diffeomorphism

$$\begin{aligned} i'_{xy}\colon \mathcal{N}'_{xy} &\to \mathcal{V}'_{xy} \\ v &\mapsto \exp_x(v + \exp_x^{-1}(y)). \end{aligned}$$

Thus $I_{xy} \triangleq I'_{xy} \circ \kappa_x^{-1}$ is a real analytic vector bundle isomorphism from $\mathsf{E}|\mathcal{U}'_{xy}$ to $\mathsf{E}|\mathcal{V}'_{xy}$ for appropriate neighbourhoods $\mathcal{U}'_{xy} \subseteq \mathsf{M}$ of x and $\mathcal{V}'_{xy} \subseteq \mathsf{M}$ of y. If we define $i_{xy}\colon \mathcal{U}'_{xy} \to \mathcal{V}'_{xy}$ by $i_{xy} = i'_{xy} \circ \exp_x^{-1}$, then I_{xy} is a vector bundle mapping over i_{xy}. Along similar lines, $\hat{I}_{xy} \triangleq \kappa_y^{-1} \circ I'_{xy}$ is a vector bundle isomorphism between the trivial bundles $\mathcal{O}'_{xy} \times \mathsf{E}_x$ and $\mathcal{N}'_{xy} \times \mathsf{E}_y$ for appropriate neighbourhoods $\mathcal{O}'_{xy} \subseteq \mathsf{T}_x\mathsf{M}$ and $\mathcal{N}'_{xy} \subseteq \mathsf{T}_y\mathsf{M}$ of the origin. If we define $\hat{\imath}_{xy}\colon \mathcal{O}'_{xy} \to \mathcal{N}'_{xy}$ by $\hat{\imath}_{xy} = \exp_y^{-1} \circ i'_{xy}$, then $\hat{I}_{xy}$ is a vector bundle map over $\hat{\imath}_{xy}$.

The next sublemma indicates that the neighbourhoods $\mathcal{U}'_{xy}$ of x and $\mathcal{O}'_{xy}$ of 0_x can be uniformly bounded from below.

Sublemma 6 *The neighbourhood $\mathcal{V}_x$ and the neighbourhoods $\mathcal{U}'_{xy}$ and $\mathcal{O}'_{xy}$ above may be chosen so that*

$$\operatorname{int}(\cap_{y\in\mathcal{V}_x}\mathcal{U}'_{xy}) \neq \emptyset, \quad \operatorname{int}(\cap_{y\in\mathcal{V}_x}\mathcal{O}'_{xy}) \neq \emptyset.$$

Proof: By [6, Theorem III.8.7] we can choose $\mathcal{V}_x$ so that, if $y \in \mathcal{V}_x$, then there is a normal coordinate neighbourhood $\mathcal{V}_y$ of y containing $\mathcal{V}_x$. Taking $\mathcal{V}'_{xy} = \mathcal{V}_x \cap \mathcal{V}_y$ and $\mathcal{U}'_{xy} = \mathcal{V}_x$ gives the sublemma. ∇

We shall always assume $\mathcal{V}_x$ chosen as in the preceding sublemma, and we let $\mathcal{U}'_x \subseteq \mathsf{M}$ be a neighbourhood of x and $\mathcal{O}'_x \subseteq \mathsf{T}_x\mathsf{M}$ be a neighbourhood of 0_x such that

$$\mathcal{U}'_x \subseteq \operatorname{int}(\cap_{y\in\mathcal{V}_x}\mathcal{U}'_{xy}), \quad \mathcal{O}'_x \subseteq \operatorname{int}(\cap_{y\in\mathcal{V}_x}\mathcal{O}'_{xy}).$$

These constructions can be "bundled together" as one to include the dependence on $y \in \mathcal{V}_x$ in a clearer manner. Since this will be useful for us, we explain it here. Let us denote $\mathcal{D}_x = \mathcal{V}_x \times \mathcal{U}'_x$, let $\mathrm{pr}_2\colon \mathcal{D}_x \to \mathcal{U}'_x$ be the projection onto the second factor, and denote

$$\begin{aligned} i_x\colon \mathcal{D}_x &\to \mathsf{M} \\ (y, x') &\mapsto i_{xy}(x'). \end{aligned}$$

Consider the pull-back bundle $\mathrm{pr}_2^*\pi\colon \mathrm{pr}_2^*\mathsf{E}|\mathcal{U}'_x \to \mathcal{D}_x$. Thus

$$\mathrm{pr}_2^*\mathsf{E}|\mathcal{U}'_x = \{((y, x'), e_{y'}) \in \mathcal{D}_x \times \mathsf{E}|\mathcal{U}'_x \mid y' = x'\}.$$

We then have a real analytic vector bundle mapping

$$\begin{aligned} I_x\colon \mathrm{pr}_2^*\mathsf{E}|\mathcal{U}'_x &\to \mathsf{E} \\ ((y, x'), e_{x'}) &\mapsto I_{xy}(e_{x'}) \end{aligned}$$

which is easily verified to be defined over i_x and is isomorphic on fibres. Given $\xi \in \Gamma^\infty(\mathsf{E})$, we define $I_x^*\xi \in \Gamma^\infty(\mathrm{pr}_2^*\mathsf{E}|\mathcal{U}'_x)$ by

$$I_x^*\xi(y, x') = (I_x)^{-1}_{(y,x')} \circ \xi \circ i_x(y, x') = I_{xy}^{-1} \circ \xi \circ i_{xy}(x').$$

For $y \in \mathcal{V}_x$ fixed, we denote by $I_{xy}^*\xi \in \Gamma^\infty(\mathsf{E}|\mathcal{U}'_x)$ the section given by

$$I_{xy}^*\xi(x') = I_x^*\xi(y, x') = I_{xy}^{-1} \circ \xi \circ i_{xy}(x').$$

A similar construction can be made in the local trivialisations. Here we denote $\hat{\mathcal{D}}_x = \mathcal{V}_x \times \mathcal{O}_x$, let $\mathrm{pr}_2\colon \hat{\mathcal{D}}_x \to \mathcal{O}_x$ be the projection onto the second factor, and consider the map

$$\begin{aligned} \hat{i}_x\colon \hat{\mathcal{D}}_x &\to \mathsf{TM} \\ (y, v_x) &\mapsto \hat{i}_{xy}(v_x). \end{aligned}$$

Denote by $\pi_{\mathsf{TM}}^*\pi\colon \pi_{\mathsf{TM}}^*\mathsf{E} \to \mathsf{TM}$ the pull-back bundle and also define the pull-back bundle

$$\mathrm{pr}_2^*\pi_{\mathsf{TM}}^*\pi\colon \mathrm{pr}_2^*\pi_{\mathsf{TM}}^*\mathsf{E} \to \hat{\mathcal{D}}_x.$$

Note that

$$\mathrm{pr}_2^*\pi_{\mathsf{TM}}^*\mathsf{E} = \{((y, v_x), (u_y, e_y)) \in \hat{\mathcal{D}}_x \times \pi_{\mathsf{TM}}^*\mathsf{E} \mid x = y\}.$$

We then define the real analytic vector bundle map

$$\begin{aligned} \hat{I}_x\colon \mathrm{pr}_2^*\pi_{\mathsf{TM}}^*\mathsf{E} &\to \pi_{\mathsf{TM}}^*\mathsf{E} \\ ((y, v_x), (u_x, e_x)) &\mapsto (v_x, \hat{I}_{xy}(v_x, e_x)). \end{aligned}$$

Given a local section $\eta \in \Gamma^\infty(\pi_{\mathsf{TM}}^*\mathsf{E})$ defined in a neighbourhood of the zero section, define a local section $\hat{I}_x^*\eta \in \Gamma^\infty(\mathrm{pr}_2^*\,\pi_{\mathsf{TM}}^*\mathsf{E})$ in a neighbourhood of the zero section of $\hat{\mathcal{D}}_x$ by

$$\hat{I}_x^*\eta(y, v_x) = (\hat{I}_x)^{-1}_{(y,v_x)} \circ \eta \circ \hat{\imath}_x(y, v_x) = \hat{I}_{xy}^{-1} \circ \eta \circ \hat{\imath}_x(y, v_x).$$

For $y \in \mathcal{V}_x$ fixed, we denote by η_y the restriction of η to a neighbourhood of $0_y \in \mathsf{T}_y\mathsf{M}$. We then denote by

$$\hat{I}_{xy}^*\eta_y(v_x) = \hat{I}_x^*\eta(y, v_x) = \hat{I}_{xy}^{-1} \circ \eta_y \circ \hat{\imath}_{xy}(v_x)$$

the element of $\Gamma^\infty(\mathcal{O}_x' \times \mathsf{E}_x)$.

The following simple lemma ties the preceding two constructions together.

Sublemma 7 *Let $\xi \in \Gamma^\infty(\mathsf{E})$ and let $\hat{\xi} \in \Gamma^\infty(\pi_{\mathsf{TM}}^*\mathsf{E})$ be defined in a neighbourhood of the zero section by*

$$\hat{\xi} = \kappa_y^{-1} \circ \xi \circ \exp_y .$$

Then, for each $y \in \mathcal{V}_x$,

$$\hat{I}_{xy}^*\hat{\xi}_y = \kappa_x^{-1} \circ I_{xy}^*\xi \circ \exp_x .$$

Proof: We have

$$\begin{aligned}
\hat{I}_{xy}^*\hat{\xi}(v_x) &= \hat{I}_{xy}^{-1} \circ \hat{\xi} \circ \hat{\imath}_{xy}(v_x) \\
&= (I_{xy}')^{-1} \circ \kappa_y \circ \hat{\xi} \circ \exp_y^{-1} \circ i_{xy}'(v_x) \\
&= \kappa_x^{-1} \circ I_{xy}^{-1} \circ \kappa_y \circ \hat{\xi} \circ \exp_y^{-1} \circ i_{xy} \circ \exp_x(v_x) \\
&= \kappa_x^{-1} \circ I_{xy}^{-1} \circ \xi \circ i_{xy} \circ \exp_x(v_x) \\
&= \kappa_x^{-1} I_{xy}^*\xi \circ \exp_x(v_x).
\end{aligned}$$

as claimed. ▽

We now use the preceding constructions to "smear" the pointwise estimate of Sublemma 5 to a neighbourhood of the fixed point. In the statement and proof of the following sublemma, we make free use of the notation we introduced in the preceding paragraphs.

Sublemma 8 *Let $\mathcal{U} \subseteq \mathbb{R}^n$ be open, denote $\mathbb{R}^k_{\mathcal{U}} = \mathcal{U} \times \mathbb{R}^k$, and consider the trivial vector bundle $\mathrm{pr}_1 : \mathbb{R}^k_{\mathcal{U}} \to \mathcal{U}$. Let $\mathbb{G}$ be a Riemannian metric on $\mathcal{U}$, let $\mathbb{G}_0$ be a vector bundle metric on $\mathbb{R}^k_{\mathcal{U}}$, let ∇ be an affine connection on $\mathcal{U}$, and let ∇^0 be a vector bundle connection on $\mathbb{R}^k_{\mathcal{U}}$, with all of these being real analytic. For each $\boldsymbol{x} \in \mathcal{U}$ there exist a neighbourhood $\mathcal{V}_{\boldsymbol{x}}$ and $C_{\boldsymbol{x}}, \sigma_{\boldsymbol{x}} \in \mathbb{R}_{>0}$ such that we have the following inequalities for each $\boldsymbol{\xi} \in \Gamma^\infty(\mathbb{R}^k_{\mathcal{U}})$, $m \in \mathbb{Z}_{\geq 0}$, and $\boldsymbol{y} \in \mathcal{V}_{\boldsymbol{x}}$:*

(i) $\sup\{\frac{1}{I!}|\boldsymbol{D}^I\hat{\xi}^a_{\boldsymbol{y}}(\boldsymbol{0})| \mid |I| \leq m,\ a \in \{1,\dots,k\}\}$
$\leq C_{\boldsymbol{x}}\sigma_{\boldsymbol{x}}^{-1} \sup\{\frac{1}{I!}|\boldsymbol{D}^I((\hat{I}_{\boldsymbol{xy}}^*)^{-1}\hat{\boldsymbol{\xi}}_{\boldsymbol{y}})^a(\boldsymbol{0})| \mid |I| \leq m,\ a \in \{1,\dots,k\}\}$;

(ii) $\sup\{\frac{1}{I!}|\boldsymbol{D}^I(\hat{I}_{\boldsymbol{xy}}^*\hat{\boldsymbol{\xi}}_{\boldsymbol{y}})^a|(\boldsymbol{0}) \mid |I| \leq m,\ a \in \{1,\dots,k\}\}$
$\leq C_{\boldsymbol{x}}\sigma_{\boldsymbol{x}}^{-1} \sup\{\frac{1}{I!}|\boldsymbol{D}^I\hat{\xi}^a_{\boldsymbol{y}}(\boldsymbol{0})| \mid |I| \leq m,\ a \in \{1,\dots,k\}\}$;

(iii) $\|j_m\xi(y)\|_{\overline{\mathbb{G}}_m} \le C_x\sigma_x^{-1}\|j_m((I^*_{xy})^{-1}\xi)(x)\|_{\overline{\mathbb{G}}_m}$;
(iv) $\|j_m(I^*_{xy}\xi)(x)\|_{\overline{\mathbb{G}}_m} \le C_x\sigma_x^{-1}\|j_m\xi(y)\|_{\overline{\mathbb{G}}_m}$.

Proof: We begin the proof with an observation. Suppose that we have an open subset $\mathcal{U} \subseteq \mathbb{R}^n\times\mathbb{R}^k$ and $f \in C^\omega(\mathcal{U})$. We wish to think of f as a function of $\boldsymbol{x} \in \mathbb{R}^n$ depending on a parameter $\boldsymbol{p} \in \mathbb{R}^k$ in a jointly real analytic manner. We note that, for $K \subseteq \mathcal{U}$ compact, we have $C, \sigma \in \mathbb{R}_{>0}$ such that the partial derivatives satisfy a bound

$$|\boldsymbol{D}_1^I f(\boldsymbol{x}, \boldsymbol{p})| \le CI!\sigma^{-|I|}$$

for every $(\boldsymbol{x}, \boldsymbol{p}) \in K$ and $I \in \mathbb{Z}^n_{\ge 0}$. This is a mere specialisation of [8, Proposition 2.2.10] to partial derivatives. The point is that the bound for the partial derivatives is uniform in the parameter $\boldsymbol{p}$. With this in mind, we note that the following are easily checked:

1. the estimate of Sublemma 1 can be extended to the case where f depends in a jointly real analytic manner on a parameter, and the estimate is uniform in the parameter over compact sets;
2. the estimate of Sublemma 2 can be extended to the case where $\boldsymbol{\Phi}$ depends in a jointly real analytic manner on a parameter, and the estimate is uniform in the parameter over compact sets;
3. as a consequence of the preceding fact, the estimate of Sublemma 3 can be extended to the case where $\boldsymbol{\Phi}$ depends in a jointly real analytic manner on a parameter, and the estimate is uniform in the parameter over compact sets;
4. as a consequence of the preceding three facts, the estimate of Sublemma 4 can be extended to the case where $\boldsymbol{\Phi}$ and $\boldsymbol{A}$ depend in a jointly real analytic manner on a parameter, and the estimate is uniform in the parameter over compact sets.

Now let us proceed with the proof.

We take $\mathcal{V}_x$ as in the discussion preceding the statement of the sublemma. Let us introduce coordinate notation for all maps needed. We have

$$\begin{aligned}
\hat{\xi}_y(\boldsymbol{u}) &= \hat{\xi}(\boldsymbol{y}, \boldsymbol{u}) = \xi \circ \exp_y(\boldsymbol{u}),\\
I^*_{xy}\xi(\boldsymbol{x}') &= \boldsymbol{A}(\boldsymbol{y}, \boldsymbol{x}') \cdot (\xi \circ i_{xy}(\boldsymbol{x}')),\\
\hat{I}^*_{xy}\hat{\xi}_y(\boldsymbol{v}) &= \hat{\boldsymbol{A}}(\boldsymbol{y}, \boldsymbol{v}) \cdot (\hat{\xi}_y \circ \hat{i}_{xy}(\boldsymbol{v})),\\
(I^*_{xy})^{-1}\xi(\boldsymbol{y}') &= \boldsymbol{A}^{-1}(\boldsymbol{y}, i_{xy}^{-1}(\boldsymbol{y}') \cdot (\xi \circ i_{xy}^{-1}(\boldsymbol{y}')),\\
(\hat{I}^*_{xy})^{-1}\hat{\xi}_y(\boldsymbol{v}) &= \hat{\boldsymbol{A}}^{-1}(\boldsymbol{y}, \hat{i}_{xy}^{-1}(\boldsymbol{u})) \cdot (\hat{\xi}_y \circ \hat{i}_{xy}^{-1}(\boldsymbol{v})),
\end{aligned}$$

for appropriate real analytic mappings $\boldsymbol{A}$ and $\hat{\boldsymbol{A}}$ taking values in $\mathsf{GL}(k;\mathbb{R})$. Note that, for every $I \in \mathbb{Z}^n_{\ge 0}$,

$$\boldsymbol{D}^I(\hat{I}^*_{xy}\hat{\xi}_y)(\boldsymbol{0}) = \boldsymbol{D}^I_2(\hat{I}^*_x\hat{\xi})(\boldsymbol{y}, \boldsymbol{0}),$$

and similarly for $\boldsymbol{D}^I((\hat{I}^*_{xy})^{-1}\hat{\xi}_y)(\boldsymbol{0})$. The observation made at the beginning of the proof shows that parts (i) and (ii) follow immediately from Sublemma 4. Parts (iii) and (iv) follow from the first two parts after an application of Sublemma 5. ▽

By applications of (a) Sublemma 8, (b) Sublemmata 7 and 5, (c) Sublemma 8 again, and (d) Sublemma 4, there exist

$$A_{1,x}, A_{2,x}, A_{3,x}, A_{4,x}, r_{1,x}, r_{2,x}, r_{3,x}, r_{4,x} \in \mathbb{R}_{>0}$$

and a relatively compact neighbourhood $\mathcal{V}_x \subseteq \mathcal{U}$ of x such that

$$\begin{aligned}
\|j_m\xi(y)\|_{\overline{\mathbb{G}}_m} &\leq A_{1,x} r_{1,x}^{-m} \|j_m((I^*_{xy})^{-1}\xi)(x)\|_{\overline{\mathbb{G}}_m} \\
&\leq A_{2,x} r_{2,x}^{-m} \sup\left\{\frac{1}{I!}|D^I((\hat{I}^*_{xy})^{-1}\hat{\xi}_y)^a(0)| \;\middle|\; |I| \leq m,\ a \in \{1,\dots,k\}\right\} \\
&\leq A_{3,x} r_{3,x}^{-m} \sup\left\{\frac{1}{I!}|D^I\hat{\xi}^a_y(0)| \;\middle|\; |I| \leq m,\ a \in \{1,\dots,k\}\right\} \\
&\leq A_{4,x} r_{4,x}^{-m} \sup\left\{\frac{1}{I!}|D^I\xi^a(y)| \;\middle|\; |I| \leq m,\ a \in \{1,\dots,k\}\right\}
\end{aligned}$$

for every $\xi \in \Gamma^\infty(\mathbb{R}^k_{\mathcal{U}})$, $m \in \mathbb{Z}_{\geq 0}$, and $y \in \mathcal{V}_x$. Take $x_1, \dots, x_k \in K$ such that $K \subseteq \cup_{j=1}^k \mathcal{V}_{x_j}$ and define

$$C_1 = \max\{A_{4,x_1}, \dots, A_{4,x_k}\}, \quad \sigma_1 = \min\{r_{4,x_1}, \dots, r_{4,x_k}\},$$

so that

$$\|j_m\xi(x)\|_{\overline{\mathbb{G}}_m} \leq C_1\sigma_1^{-m} \sup\left\{\frac{1}{I!}|D^I\xi^a(x)| \;\middle|\; |I| \leq m,\ a \in \{1,\dots,k\}\right\}$$

for every $\xi \in \Gamma^\infty(\mathbb{R}^k_{\mathcal{U}})$, $m \in \mathbb{Z}_{\geq 0}$, and $x \in K$. This gives one half of the estimate in the lemma.

For the other half of the estimate in the lemma, we apply (a) Sublemma 4, (b) Sublemma 8, (c) Sublemmata 7 and 5, and (d) Sublemma 8 again to assert the existence of

$$A_{1,x}, A_{2,x}, A_{3,x}, A_{4,x}, r_{1,x}, r_{2,x}, r_{3,x}, r_{4,x} \in \mathbb{R}_{>0}$$

and a relatively compact neighbourhood $\mathcal{V}_x \subseteq \mathcal{U}$ of x such that

$$\begin{aligned}
&\sup\left\{\frac{1}{I!}|D^I\xi^a(y)| \;\middle|\; |I| \leq m,\ a \in \{1,\dots,m\}\right\} \\
&\quad\leq A_{1,x} r_{1,x}^{-m} \sup\left\{\frac{1}{I!}|D^I\hat{\xi}^a_y(0)| \;\middle|\; |I| \leq m,\ a \in \{1,\dots,k\}\right\} \\
&\quad\leq A_{2,x} r_{2,x}^{-m} \sup\left\{\frac{1}{I!}|D^I((\hat{I}^*_{xy})^{-1}\hat{\xi}_y)^a(0)| \;\middle|\; |I| \leq m,\ a \in \{1,\dots,k\}\right\} \\
&\quad\leq A_{3,x} r_{3,x}^{-m} \|j_m((I^*_{xy})^{-1}\xi)(x)\|_{\overline{\mathbb{G}}_m} \leq A_{4,x} r_{4,x}^{-m} \|j_m\xi(y)\|_{\overline{\mathbb{G}}_m}
\end{aligned}$$

for every $\xi \in \Gamma^\infty(\mathbb{R}^k_{\mathcal{U}})$, $m \in \mathbb{Z}_{\geq 0}$, and $y \in \mathcal{V}_x$. As we argued above using a standard compactness argument, there exist $C_2, \sigma_2 \in \mathbb{R}_{>0}$ such that

$$\sup\left\{\frac{1}{I!}|D^I\xi^a(x)| \;\middle|\; |I| \leq m,\ a \in \{1,\dots,k\}\right\} \leq C_2\sigma_2^{-m}\|j_m\xi(x)\|_{\overline{\mathbb{G}}_m}$$

for every $\boldsymbol{\xi} \in \Gamma^\infty(\mathbb{R}^k_{\mathcal{U}})$, $m \in \mathbb{Z}_{\geq 0}$, and $\boldsymbol{x} \in K$. Taking $C = \max\{C_1, C_2\}$ and $\sigma = \min\{\sigma_1, \sigma_2\}$ gives the lemma. □

The preceding lemma will come in handy on a few crucial occasions. To illustrate how it can be used, we give the following characterisation of real analytic sections, referring to Chap. 3 for the definition of the seminorm $p^\infty_{K,m}$ used in the statement.

Lemma 2.6 (Characterisation of Real Analytic Sections). *Let $\pi\colon \mathsf{E} \to \mathsf{M}$ be a real analytic vector bundle and let $\xi \in \Gamma^\infty(\mathsf{E})$. Then the following statements are equivalent:*

(i) $\xi \in \Gamma^\omega(\mathsf{E})$;

(ii) for every compact set $K \subseteq \mathsf{M}$, there exist $C, r \in \mathbb{R}_{>0}$ such that $p^\infty_{K,m}(\xi) \leq Cr^{-m}$ for every $m \in \mathbb{Z}_{\geq 0}$.

Proof: (i) $\implies$ (ii) Let $K \subseteq \mathsf{M}$ be compact, let $x \in K$, and let $(\mathcal{V}_x, \psi_x)$ be a vector bundle chart for E with $(\mathcal{U}_x, \phi_x)$ the corresponding chart for M. Let $\boldsymbol{\xi}\colon \phi(\mathcal{U}_x) \to \mathbb{R}^k$ be the local representative of ξ. By [8, Proposition 2.2.10], there exist a neighbourhood $\mathcal{U}'_x \subseteq \mathcal{U}_x$ of x and $B_x, \sigma_x \in \mathbb{R}_{>0}$ such that

$$|\boldsymbol{D}^I \xi^a(\boldsymbol{x}')| \leq B_x I! \sigma_x^{-|I|}$$

for every $a \in \{1, \ldots, k\}$, $\boldsymbol{x}' \in \mathrm{cl}(\mathcal{U}'_x)$, and $I \in \mathbb{Z}^n_{\geq 0}$. We can suppose, without loss of generality, that $\sigma_x \in (0, 1)$. In this case, if $|I| \leq m$,

$$\frac{1}{I!}|\boldsymbol{D}^I \xi^a(\boldsymbol{x}')| \leq B_x \sigma_x^{-m}$$

for every $a \in \{1, \ldots, k\}$ and $\boldsymbol{x}' \in \mathrm{cl}(\mathcal{U}'_x)$. By Lemma 2.5, there exist $C_x, r_x \in \mathbb{R}_{>0}$ such that

$$\|j_m\xi(x')\|_{\overline{\mathbb{G}}_m} \leq C_x r_x^{-m}, \qquad x' \in \mathrm{cl}(\mathcal{U}'_x),\ m \in \mathbb{Z}_{\geq 0}.$$

Let $x_1, \ldots, x_k \in K$ be such that $K \subseteq \cup_{j=1}^k \mathcal{U}'_{x_j}$ and let $C = \max\{C_{x_1}, \ldots, C_{x_k}\}$ and $r = \min\{r_{x_1}, \ldots, r_{x_k}\}$. Then, if $x \in K$, we have $x \in \mathcal{U}'_{x_j}$ for some $j \in \{1, \ldots, k\}$ and so

$$\|j_m\xi(x)\|_{\overline{\mathbb{G}}_m} \leq C_{x_j} r_{x_j}^{-m} \leq Cr^{-m},$$

as desired.

(ii) $\implies$ (ii) Let $x \in \mathsf{M}$ and let $(\mathcal{V}, \psi)$ be a vector bundle chart for E such that the associated chart $(\mathcal{U}, \phi)$ for M is a relatively compact coordinate chart about x. Let $\boldsymbol{\xi}\colon \phi(\mathcal{U}) \to \mathbb{R}^k$ be the local representative of ξ. By hypothesis, there exist $C, r \in \mathbb{R}_{>0}$ such that $\|j_m\xi(x')\|_{\overline{\mathbb{G}}_m} \leq Cr^{-m}$ for every $m \in \mathbb{Z}_{\geq 0}$ and $x' \in \mathcal{U}$. Let $\mathcal{U}'$ be a relatively compact neighbourhood of x such that $\mathrm{cl}(\mathcal{U}') \subseteq \mathcal{U}$. By Lemma 2.5, there exist $B, \sigma \in \mathbb{R}_{>0}$ such that

$$|\boldsymbol{D}^I \xi^a(\boldsymbol{x}')| \leq BI! \sigma^{-|I|}$$

for every $a \in \{1, \ldots, k\}$, $\boldsymbol{x}' \in \mathrm{cl}(\mathcal{U}')$, and $I \in \mathbb{Z}^n_{\geq 0}$. We conclude real analyticity of ξ in a neighbourhood of x by [8, Proposition 2.2.10]. □

References

1. Abraham, R., Marsden, J.E., Ratiu, T.S.: Manifolds, Tensor Analysis, and Applications, 2 edn. No. 75 in Applied Mathematical Sciences. Springer-Verlag (1988)
2. Constantine, G.M., Savits, T.H.: A multivariate Faà di Bruno formula with applications. Transactions of the American Mathematical Society **348**(2), 503–520 (1996)
3. Goldschmidt, H.L.: Existence theorems for analytic linear partial differential equations. Annals of Mathematics. Second Series **86**(2), 246–270 (1967)
4. Grauert, H.: On Levi's problem and the imbedding of real-analytic manifolds. Annals of Mathematics. Second Series **68**, 460–472 (1958)
5. Halmos, P.R.: Finite-Dimensional Vector Spaces, 2 edn. Undergraduate Texts in Mathematics. Springer-Verlag, New York/Heidelberg/Berlin (1986)
6. Kobayashi, S., Nomizu, K.: Foundations of Differential Geometry, Volume I. No. 15 in Interscience Tracts in Pure and Applied Mathematics. Interscience Publishers, New York (1963)
7. Kolář, I., Michor, P.W., Slovák, J.: Natural Operations in Differential Geometry. Springer-Verlag, New York/Heidelberg/Berlin (1993)
8. Krantz, S.G., Parks, H.R.: A Primer of Real Analytic Functions, 2 edn. Birkhäuser Advanced Texts. Birkhäuser, Boston/Basel/Stuttgart (2002)
9. Kriegl, A., Michor, P.W.: The Convenient Setting of Global Analysis. No. 57 in American Mathematical Society Mathematical Surveys and Monographs. American Mathematical Society, Providence, RI (1997)
10. Pohl, W.F.: Connexions in differential geometry of higher order. Transactions of the American Mathematical Society **125**(2), 310–325 (1966)
11. Saunders, D.J.: The Geometry of Jet Bundles. No. 142 in London Mathematical Society Lecture Note Series. Cambridge University Press, New York/Port Chester/Melbourne/Sydney (1989)
12. Sontag, E.D.: Mathematical Control Theory: Deterministic Finite Dimensional Systems, 2 edn. No. 6 in Texts in Applied Mathematics. Springer-Verlag, New York/Heidelberg/Berlin (1998)
13. Thilliez, V.: Sur les fonctions composées ultradifférentiables. Journal de Mathématiques Pures et Appliquées. Neuvième Sér **76**, 499–524 (1997)

Chapter 3
The Compact-Open Topologies for the Spaces of Finitely Differentiable, Lipschitz, and Smooth Vector Fields

As mentioned in Chap. 1, one of the themes of this work is the connection between topologies for spaces of vector fields and regularity of their flows. In this and the subsequent two chapters we describe appropriate topologies for finitely differentiable, Lipschitz, smooth, holomorphic, and real analytic vector fields. The topology we use in this chapter in the smooth case (and the easily deduced finitely differentiable case) is classical, and is described, for example, in [2, §2.2]; see also [16, Chapter 4]. What we do that is original is provide a characterisation of the seminorms for this topology using the jet bundle fibre metrics from Sect. 2.2. The fruits of the effort expended in the next three chapters is harvested when our concrete definitions of seminorms permit a relatively unified analysis in Chap. 6 of time-varying vector fields.

One facet of our presentation that is novel is that we flesh out completely the "weak-$\mathscr{L}$" characterisations of topologies for vector fields. These topologies characterise vector fields by how they act on functions through Lie differentiation. The use of such "weak" characterisations is commonplace, e.g., [2, 25], although the equivalence with strong characterisation is not typically proved; indeed, we know of no existing proofs of our Theorems 3.5, 3.8, 3.14, 4.5, and 5.8. We show that, for the issues that come up in this monograph, the weak characterisations for vector field topologies agree with the direct "strong" characterisations. This requires some detailed knowledge of the topologies we use.

While our primary interest is in vector fields, i.e., sections of the tangent bundle, it is advantageous to work instead with topologies for sections of general vector bundles, and then specialise to vector fields. We will also work with topologies for functions, but this falls out easily from the general vector bundle treatment.

S. Jafarpour, A.D. Lewis, *Time-Varying Vector Fields and Their Flows*,
SpringerBriefs in Mathematics, DOI 10.1007/978-3-319-10139-2_3

3.1 General Smooth Vector Bundles

We let $\pi\colon \mathsf{E} \to \mathsf{M}$ be a smooth vector bundle with ∇^0 a linear connection on E, ∇ an affine connection on M, $\mathbb{G}_0$ a fibre metric on E, and $\mathbb{G}$ a Riemannian metric on M. This gives us, as in Sect. 2.2, fibre metrics $\overline{\mathbb{G}}_m$ on the jet bundles $\mathsf{J}^m\mathsf{E}$, $m \in \mathbb{Z}_{\geq 0}$, and corresponding fibre norms $\|\cdot\|_{\overline{\mathbb{G}}_m}$.

For a compact set $K \subseteq \mathsf{M}$ and for $m \in \mathbb{Z}_{\geq 0}$, we now define a seminorm $p^{\infty}_{K,m}$ on $\Gamma^{\infty}(\mathsf{E})$ by

$$p^{\infty}_{K,m}(\xi) = \sup\{\|j_m\xi(x)\|_{\overline{\mathbb{G}}_m} \mid x \in K\}.$$

The locally convex topology on $\Gamma^{\infty}(\mathsf{TM})$ defined by the family of seminorms $p^{\infty}_{K,m}$, $K \subseteq \mathsf{M}$ compact, $m \in \mathbb{Z}_{\geq 0}$, is called the ***smooth compact open*** or **CO^{∞}*-topology*** for $\Gamma^{\infty}(\mathsf{E})$.

We comment that the seminorms depend on the choices of ∇, ∇^0, $\mathbb{G}$, and $\mathbb{G}_0$, but the CO^{∞}-topology is independent of these choices. We will constantly use these seminorms throughout the monograph, and in doing so we will automatically be assuming that we have selected the linear connection ∇^0, the affine connection ∇, the fibre metric $\mathbb{G}_0$, and the Riemannian metric $\mathbb{G}$. We will do this often without explicit mention of these objects having been chosen.

3.2 Properties of the CO^{∞}-Topology

Let us say a few words about the CO^{∞}-topology, referring to references for details. The locally convex CO^{∞}-topology has the following attributes.

CO^{∞}-1. It is Hausdorff: [16, 4.3.1].

CO^{∞}-2. It is complete: [16, 4.3.2].

CO^{∞}-3. It is metrisable: [16, 4.3.1].

CO^{∞}-4. It is separable: We could not find this stated anywhere, but here's a sketch of a proof. By embedding E in Euclidean space $\mathbb{R}^N$ and, using an argument like that for real analytic vector bundles in the proof of Lemma 2.4, we regard E as a subbundle of a trivial bundle over the submanifold $\mathsf{M} \subseteq \mathbb{R}^N$. In this case, we can reduce our claim of separability of the CO^{∞}-topology to that for smooth functions on submanifolds of $\mathbb{R}^N$. Here we can argue as follows. If $K \subseteq \mathsf{M}$ is compact, it can be contained in a compact cube C in $\mathbb{R}^N$. Then we can use a cutoff function to take any smooth function on M and leave it untouched on a neighbourhood of K, but have it and all of its derivatives vanish outside a compact set contained in $\mathrm{int}(C)$. Then we can use Fourier series to approximate in the CO^{∞}-topology [24, Theorem VII.2.11(b)]. Since there are countably many Fourier basis functions, this gives the desired separability.

CO^{∞}-5. It is nuclear: [12, Theorem 21.6.6].

CO$^\infty$-6. It is Suslin[1]: This follows since $\Gamma^\infty(\mathsf{TM})$ is a Polish space (see footnote 1), as we have already seen.

Some of these attributes perhaps seem obscure, but we will, in fact, use all of them!

Since the CO$^\infty$-topology is metrisable, it is exactly characterised by its convergent sequences, so let us describe these. A sequence $(\xi_k)_{k\in\mathbb{Z}_{>0}}$ in $\Gamma^\infty(\mathsf{E})$ converges to $\xi \in \Gamma^\infty(\mathsf{E})$ if and only if, for each compact set $K \subseteq \mathsf{M}$ and for each $m \in \mathbb{Z}_{\geq 0}$, the sequence $(j_m\xi_k|K)_{k\in\mathbb{Z}_{>0}}$ converges uniformly to $j_m\xi|K$, cf. combining [17, Theorem 46.8] and [16, Lemma 4.2].

Since the topology is nuclear, it follows that subsets of $\Gamma^\infty(\mathsf{TM})$ are compact if and only if they are closed and von Neumann bounded [18, Proposition 4.47]. That is to say, in a nuclear locally convex space, the compact bornology and the von Neumann bornology agree, according to the terminology introduced in Sect. 1.3. It is then interesting to characterise von Neumann bounded subsets of $\Gamma^\infty(\mathsf{E})$. One can show that a subset $\mathcal{B}$ is bounded in the von Neumann bornology if and only if every continuous seminorm on V is a bounded function when restricted to $\mathcal{B}$ [20, Theorem 1.37(b)]. Therefore, to characterise von Neumann bounded subsets, we need only characterise subsets on which each of the seminorms $p^\infty_{K,m}$ is a bounded function. This obviously gives the following characterisation.

Lemma 3.1 (Bounded Subsets in the CO$^\infty$-Topology). *A subset $\mathcal{B} \subseteq \Gamma^\infty(\mathsf{E})$ is bounded in the von Neumann bornology if and only if the following property holds: for any compact set $K \subseteq \mathsf{M}$ and any $m \in \mathbb{Z}_{\geq 0}$, there exists $C \in \mathbb{R}_{>0}$ such that $p^\infty_{K,m}(\xi) \leq C$ for every $\xi \in \mathcal{B}$.*

Let us give a coordinate characterisation of the smooth compact-open topology, just for concreteness so that the reader can see that our constructions agree with perhaps more familiar things. If we have a smooth vector bundle $\pi\colon \mathsf{E} \to \mathsf{M}$, we let $(\mathcal{V}, \psi)$ be a vector bundle chart for E inducing a chart $(\mathcal{U}, \phi)$ for M. For $\xi \in \Gamma^\infty(\mathsf{E})$, the local representative of ξ has the form

$$\mathbb{R}^n \supseteq \phi(\mathcal{U}) \ni \boldsymbol{x} \mapsto (\boldsymbol{x}, \boldsymbol{\xi}(\boldsymbol{x})) \in \phi(\mathcal{U}) \times \mathbb{R}^k.$$

Thus we have an associated map $\boldsymbol{\xi}\colon \phi(\mathcal{U}) \to \mathbb{R}^k$ that describes the section locally. A **CO$^\infty$-*subbasic neighbourhood*** is a subset $\mathcal{B}_\infty(\xi, \mathcal{V}, K, \epsilon, m)$ of $\Gamma^\infty(\mathsf{E})$, where

1. $\xi \in \Gamma^\infty(\mathsf{E})$,
2. $(\mathcal{V}, \psi)$ is a vector bundle chart for E with associated chart $(\mathcal{U}, \phi)$ for M,
3. $K \subseteq \mathcal{U}$ is compact,
4. $\epsilon \in \mathbb{R}_{>0}$,
5. $m \in \mathbb{Z}_{\geq 0}$, and
6. $\eta \in \mathcal{B}_\infty(\xi, \mathcal{V}, K, \epsilon, m)$ if and only if

$$\|\boldsymbol{D}^l\boldsymbol{\eta}(\boldsymbol{x}) - \boldsymbol{D}^l\boldsymbol{\xi}(\boldsymbol{x})\| < \epsilon, \qquad \boldsymbol{x} \in \phi(K),\ l \in \{0, 1, \dots, m\},$$

where $\boldsymbol{\xi}, \boldsymbol{\eta}\colon \phi(\mathcal{U}) \to \mathbb{R}^k$ are the local representatives.

[1] A ***Polish space*** is a complete separable metrisable space. A ***Suslin space*** is a continuous image of a Polish space. A good reference for the basic properties of Suslin spaces is [4, Chapter 6].

One can show that the CO^∞-topology is that topology having as a subbase the CO^∞-subbasic neighbourhoods. This is the definition used in [10], for example. To show that this topology agrees with our intrinsic characterisation is a straightforward bookkeeping chore, and the interested reader can refer to Lemma 2.5 to see how this is done in the more difficult real analytic case. This more concrete characterisation using vector bundle charts can be useful should one ever wish to verify some properties in examples. It can also be useful in general arguments in emergencies when one does not have the time to flesh out coordinate-free constructions.

3.3 The Weak-$\mathscr{L}$ Topology for Smooth Vector Fields

The CO^∞-topology for smooth sections of a vector bundle, merely by specialisation, gives a locally convex topology on the set $\Gamma^\infty(\mathsf{TM})$ of smooth vector fields and the set $\mathrm{C}^\infty(\mathsf{M})$ of smooth functions (noting that a smooth function is obviously identified with a section of the trivial vector bundle $\mathsf{M} \times \mathbb{R}$). The only mildly interesting thing in these cases is that one does not need a separate linear connection in the vector bundles or a separate fibre metric. Indeed, TM is already assumed to have a linear connection (the affine connection on M) and a fibre metric (the Riemannian metric on M), and the trivial bundle has the canonical flat linear connection defined by $\nabla_X f = \mathscr{L}_X f$ and the standard fibre metric induced by absolute value on the fibres.

We wish to see another way of describing the CO^∞-topology on $\Gamma^\infty(\mathsf{TM})$ by noting that a vector field defines a linear map, indeed a derivation, on $\mathrm{C}^\infty(\mathsf{M})$ by Lie differentiation: $f \mapsto \mathscr{L}_X f$. The topology we describe for $\Gamma^\infty(\mathsf{TM})$ is a sort of weak topology arising from the CO^∞-topology on $\mathrm{C}^\infty(\mathsf{M})$ and Lie differentiation. To properly set the stage for the fact that we will repeat this construction for our other topologies, it is most clear to work in a general setting for a moment and then specialise in each subsequent case.

The general setup is provided by the next definition.

Definition 3.2 (Weak Boundedness, Continuity, Measurability, and Integrability). Let $\mathbb{F} \in \{\mathbb{R}, \mathbb{C}\}$ and let U and V be $\mathbb{F}$-vector spaces with V locally convex. Let $\mathscr{A} \subseteq \mathrm{Hom}_{\mathbb{R}}(\mathsf{U}; \mathsf{V})$ and let the ***weak-$\mathscr{A}$ topology*** on U be the weakest topology for which A is continuous for every $A \in \mathscr{A}$ [11, §2.11].

Also let $(\mathcal{X}, \mathscr{O})$ be a topological space, let $(\mathcal{T}, \mathscr{M})$ be a measurable space, and let $\mu\colon \mathscr{M} \to \overline{\mathbb{R}}_{\geq 0}$ be a finite measure. We have the following notions:

(i) a subset $\mathcal{B} \subseteq \mathsf{U}$ is ***weak-$\mathscr{A}$ bounded in the von Neumann bornology*** if $A(\mathcal{B})$ is bounded in the von Neumann bornology for every $A \in \mathscr{A}$;

(ii) a map $\Phi\colon \mathcal{X} \to \mathsf{U}$ is ***weak-$\mathscr{A}$ continuous*** if $A \circ \Phi$ is continuous for every $A \in \mathscr{A}$;

(iii) a map $\Psi\colon \mathcal{T} \to \mathsf{U}$ is ***weak-$\mathscr{A}$ measurable*** if $A \circ \Psi$ is measurable for every $A \in \mathscr{A}$;

(iv) a map $\Psi\colon \mathcal{T} \to \mathsf{U}$ is ***weak-$\mathscr{A}$ Bochner integrable*** with respect to μ if $A \circ \Psi$ is Bochner integrable with respect to μ for every $A \in \mathscr{A}$. ○

As can be seen in Sect. 2.11 of [11], the weak-$\mathscr{A}$ topology is a locally convex topology, and a subbase for open sets in this topology is

$$\{A^{-1}(\mathcal{O}) \mid A \in \mathscr{A},\ \mathcal{O} \subseteq \mathsf{V} \text{ open}\}.$$

Equivalently, the weak-$\mathscr{A}$ topology is defined by the seminorms

$$u \mapsto q(A(u)), \qquad A \in \mathscr{A},\ q \text{ a continuous seminorm for } \mathsf{V}.$$

This is a characterisation of the weak-$\mathscr{A}$ topology we will use often.

We now have the following result which gives conditions for the equivalence of "weak-$\mathscr{A}$" notions with the usual notions. We call a subset $\mathscr{A} \subseteq \mathrm{Hom}_{\mathbb{F}}(\mathsf{U};\mathsf{V})$ ***point separating*** if, given distinct $u_1, u_2 \in \mathsf{U}$, there exists $A \in \mathscr{A}$ such that $A(u_1) \neq A(u_2)$.

Lemma 3.3 (Equivalence of Weak-$\mathscr{A}$ and Locally Convex Notions for General Locally Convex Spaces). *Let $\mathbb{F} \in \{\mathbb{R}, \mathbb{C}\}$ and let U and V be locally convex $\mathbb{F}$-vector spaces. Let $\mathscr{A} \subseteq \mathrm{Hom}_{\mathbb{R}}(\mathsf{U};\mathsf{V})$ and suppose that the weak-$\mathscr{A}$ topology agrees with the locally convex topology for U. Let $(\mathcal{X}, \mathscr{O})$ be a topological space, let $(\mathcal{T}, \mathscr{M})$ be a measurable space, and let $\mu\colon \mathscr{M} \to \overline{\mathbb{R}}_{\geq 0}$ be a finite measure. Then the following statements hold:*

(i) a subset $\mathcal{B} \subseteq \mathsf{U}$ is bounded in the von Neumann bornology if and only if it is weak-$\mathscr{A}$ bounded in the von Neumann bornology;

(ii) a map $\Phi\colon \mathcal{X} \to \mathsf{U}$ is continuous if and only if it is weak-$\mathscr{A}$ continuous;

(iii) for a map $\Psi\colon \mathcal{T} \to \mathsf{U}$,

(a) if Ψ is measurable, then it is weak-$\mathscr{A}$ measurable;

(b) if U and V are Hausdorff Suslin spaces, if $\mathscr{A}$ contains a countable point separating subset, and if Ψ is weak-$\mathscr{A}$ measurable, then Ψ is measurable;

(iv) if U is complete and separable, a map $\Psi\colon \mathcal{T} \to \mathsf{U}$ is Bochner integrable with respect to μ if and only if it is weak-$\mathscr{A}$ Bochner integrable with respect to μ.

Proof: (i) and (ii): Both of these assertions follow directly from the fact that the locally convex topology of U agrees with the weak-$\mathscr{A}$ topology. Indeed, the equivalence of these topologies implies that (a) if p is a continuous seminorm for the locally convex topology of U, then there exist continuous seminorms $q_1, \dots, q_k$ for V and $A_1, \dots, A_k \in \mathscr{A}$ such that

$$p(u) \leq q_1(A_1(u)) + \cdots + q_k(A_k(u)), \qquad u \in \mathsf{U}, \tag{3.1}$$

and (b) if q is a continuous seminorm for V and if $A \in \mathscr{A}$, then there exists a continuous seminorm p for the locally convex topology for U such that

$$q(A(u)) \leq p(u), \qquad u \in \mathsf{U}. \tag{3.2}$$

(iii) First suppose that Ψ is measurable and let $A \in \mathscr{A}$. Since the locally convex topology of U agrees with the weak-$\mathscr{A}$ topology, A is continuous in the locally convex topology of U. Therefore, if Ψ is measurable, it follows immediately by continuity of A that $A \circ \Psi$ is measurable.

Next suppose that U and V are Suslin, that $\mathscr{A}$ contains a countable point separating subset, and that Ψ is weak-$\mathscr{A}$ measurable. Without loss of generality, let us suppose that $\mathscr{A}$ is itself countable. By $\mathsf{V}^{\mathscr{A}}$ we denote the mappings from $\mathscr{A}$ to V, with the usual pointwise vector space structure. A typical element of $\mathsf{V}^{\mathscr{A}}$ we denote by ϕ. By [4, Lemma 6.6.5(iii)], $\mathsf{V}^{\mathscr{A}}$ is a Suslin space. Let us define a mapping $\iota_{\mathscr{A}}\colon \mathsf{U} \to \mathsf{V}^{\mathscr{A}}$ by $\iota_{\mathscr{A}}(u)(A) = A(u)$. Since $\mathscr{A}$ is point separating, we easily verify that $\iota_{\mathscr{A}}$ is injective, and so we have U as a subspace of the countable product $\mathsf{V}^{\mathscr{A}}$. For $A \in \mathscr{A}$ let $\mathrm{pr}_A\colon \mathsf{V}^{\mathscr{A}} \to \mathsf{V}$ be the projection defined by $\mathrm{pr}_A(\phi) = \phi(A)$. Since V is Suslin, it is hereditary Lindelöf [4, Lemma 6.6.4]. Thus the Borel σ-algebra of $\mathsf{V}^{\mathscr{A}}$ is the same as the initial Borel σ-algebra defined by the projections pr_A, $A \in \mathscr{A}$, i.e., the smallest σ-algebra for which the projections are measurable [4, Lemma 6.4.2]. By hypothesis, $(A \circ \Psi)^{-1}(\mathcal{B})$ is measurable for every $A \in \mathscr{A}$ and every Borel set $\mathcal{B} \subseteq \mathsf{V}$. Now we note that $\mathrm{pr}_A \circ \iota_{\mathscr{A}}(v) = A(v)$, from which we deduce that

$$(A \circ \Psi)^{-1}(\mathcal{B}) = (\iota_{\mathscr{A}} \circ \Psi)^{-1}(\mathrm{pr}_A^{-1}(\mathcal{B}))$$

is measurable for every $A \in \mathscr{A}$ and every Borel set $\mathcal{B} \subseteq \mathsf{V}$. Thus $\iota_{\mathscr{A}} \circ \Psi$ is measurable. Since $\iota_{\mathscr{A}}$ is continuous, the subspace topology of U induced by $\iota_{\mathscr{A}}$ is coarser than the given topology for U. Since the given topology for U is Hausdorff and Suslin, as is shown in [23], a subset $\mathcal{C} \subseteq \mathcal{U}$ is Borel if and only if $\iota_{\mathscr{A}}(\mathcal{C})$ is Borel. Thus, if $\mathcal{C} \subseteq \mathsf{U}$ is Borel, then $\Psi^{-1}(\mathcal{C}) = (\iota_{\mathscr{A}} \circ \Psi)^{-1}(\iota_{\mathscr{A}}(\mathcal{C}))$ is measurable, and this gives the desired measurability of Ψ.

(iv) Since U is separable and complete, by [3, Theorems 3.2 and 3.3] Bochner integrability of Ψ is equivalent to integrability, in the sense of Lebesgue, of $t \mapsto p \circ \Psi(t)$ for any continuous seminorm p. Thus, Ψ is Bochner integrable with respect to the locally convex topology of U if and only if $t \mapsto p \circ \Psi(t)$ is integrable, and Ψ is weak-$\mathscr{A}$ Bochner integrable if and only if $t \mapsto q_A(\Psi(t))$ is integrable for every $A \in \mathscr{A}$. This part of the proof now follows from the inequalities (3.1) and (3.2) that characterise the equivalence of the locally convex and weak-$\mathscr{A}$ topologies for U. □

The proof of the harder direction in part (iii) is an adaptation of [26, Theorem 1] to our more general setting. We will revisit this idea again when we talk about measurability of time-varying vector fields in Chap. 6.

For $f \in \mathrm{C}^\infty(\mathsf{M})$, let us define

$$\begin{aligned} \mathscr{L}_f\colon \Gamma^\infty(\mathsf{TM}) &\to \mathrm{C}^\infty(\mathsf{M}) \\ X &\mapsto \mathscr{L}_X f. \end{aligned}$$

The topology for $\Gamma^\infty(\mathsf{TM})$ we now define corresponds to the general case of Definition 3.2 by taking $\mathsf{U} = \Gamma^\infty(\mathsf{TM})$, $\mathsf{V} = \mathrm{C}^\infty(\mathsf{M})$, and $\mathscr{A} = \{\mathscr{L}_f \mid f \in \mathrm{C}^\infty(\mathsf{M})\}$. To this end, we make the following definition.

Definition 3.4 (Weak-$\mathscr{L}$ Topology for Space of Smooth Vector Fields). For a smooth manifold M, the ***weak-$\mathscr{L}$ topology*** for $\Gamma^\infty(\mathsf{TM})$ is the weakest topology for which $\mathscr{L}_f$ is continuous for every $f \in \mathrm{C}^\infty(\mathsf{M})$, if $\mathrm{C}^\infty(\mathsf{M})$ has the CO^∞-topology. ∘

We now have the following result.

Theorem 3.5 (Weak-$\mathscr{L}$ characterisation of CO$^\infty$-Topology for Smooth Vector Fields). *For a smooth manifold, the following topologies for $\Gamma^\infty(\mathsf{TM})$ agree:*
(i) the CO$^\infty$*-topology;*
(ii) the weak-$\mathscr{L}$ topology.

Proof: (i)⊆(ii) For this part of the proof, we assume that M has a well-defined dimension. The proof is easily modified by additional notation to cover the case where this may not hold. Let $K \subseteq \mathsf{M}$ be compact and let $m \in \mathbb{Z}_{\geq 0}$. Let $x \in K$ and let $(\mathcal{U}_x, \phi_x)$ be a coordinate chart for M about x with coordinates denoted by $(x^1, \dots, x^n)$. Let $\boldsymbol{X}\colon \phi_x(\mathcal{U}_x) \to \mathbb{R}^n$ be the local representative of $X \in \Gamma^\infty(\mathsf{TM})$. For $j \in \{1, \dots, n\}$ let $f_x^j \in \mathrm{C}^\infty(\mathsf{M})$ have the property that, for some relatively compact neighbourhood $\mathcal{V}_x$ of x with $\mathrm{cl}(\mathcal{V}_x) \subseteq \mathcal{U}_x$, $f_x^j = x^j$ in some neighbourhood of $\mathrm{cl}(\mathcal{V}_x)$. (This is done using standard extension arguments for smooth functions, cf. [1, Proposition 5.5.8].) Then, in a neighbourhood of $\mathrm{cl}(\mathcal{V}_x)$ in $\mathcal{U}_x$, we have $\mathscr{L}_X f_x^j = X^j$. Therefore, for each $y \in \mathrm{cl}(\mathcal{V}_x)$,

$$j_m X(y) \mapsto \sum_{j=1}^{n} \|j_m(\mathscr{L}_X f_x^j)(y)\|_{\overline{\mathbb{G}}_m}$$

is a norm on the fibre $\mathsf{J}_y^m\mathsf{E}$. Therefore, there exists $C_x \in \mathbb{R}_{>0}$ such that

$$\|j_m X(y)\|_{\overline{\mathbb{G}}_m} \leq C_x \sum_{j=1}^{n} \|j_m(\mathscr{L}_X f_x^j)(y)\|_{\overline{\mathbb{G}}_m}, \qquad y \in \mathrm{cl}(\mathcal{V}_x).$$

Since K is compact, let $x_1, \dots, x_k \in K$ be such that $K \subseteq \cup_{a=1}^k \mathcal{V}_{x_a}$. Let

$$C = \max\{C_{x_1}, \dots, C_{x_k}\}.$$

Then, if $y \in K$ we have $y \in \mathcal{V}_{x_a}$ for some $a \in \{1, \dots, k\}$, and so

$$\|j_m X(y)\|_{\overline{\mathbb{G}}_m} \leq C \sum_{j=1}^{n} \|j_m(\mathscr{L}_X f_{x_a}^j)(y)\|_{\overline{\mathbb{G}}_m} \leq C \sum_{a=1}^{k} \sum_{j=1}^{n} \|j_m(\mathscr{L}_X f_{x_a}^j)(y)\|_{\overline{\mathbb{G}}_m}.$$

Taking supremums over $y \in K$ gives

$$p_{K,m}^\infty(X) \leq C \sum_{a=1}^{k} \sum_{j=1}^{n} p_{K,m}^\infty(\mathscr{L}_X f_{x_a}^j),$$

This part of the theorem then follows since the weak-$\mathscr{L}$ topology, as we indicated following Definition 3.2 above, is defined by the seminorms

$$X \mapsto p_{K,m}^\infty(\mathscr{L}_X f), \qquad K \subseteq \mathsf{M} \text{ compact}, \ m \in \mathbb{Z}_{\geq 0}, \ f \in \mathrm{C}^\infty(\mathsf{M}).$$

(ii)⊆(i) As per (2.1), let us abbreviate

$$\nabla^j(\dots(\nabla^1(\nabla^0 A))) = \nabla^{(j)} A,$$

where A can be either a vector field or one-form, in what we will need. Since covariant differentials commute with contractions [7, Theorem 7.03(F)], an elementary induction argument gives the formula

$$\nabla^{(m-1)}(\mathrm{d}f(X)) = \sum_{j=0}^{m} \binom{m}{j} C_{1,m-j+1}((\nabla^{(m-j-1)}X) \otimes (\nabla^{(j-1)}\mathrm{d}f)), \tag{3.3}$$

where $C_{1,m-j+1}$ is the contraction defined by

$$C_{1,m-j+1}(v \otimes \alpha^1 \otimes \cdots \otimes \alpha^{m-j} \otimes \alpha^{m-j+1} \otimes \alpha^{m-j+2} \otimes \cdots \otimes \alpha^{m+1})$$
$$= (\alpha^{m-j+1}(v))(\alpha^1 \otimes \cdots \otimes \alpha^{m-j} \otimes \alpha^{m-j+2} \otimes \cdots \otimes \alpha^{m+1}).$$

In writing (3.3) we use the convention $\nabla^{(-1)}X = X$ and $\nabla^{(-1)}(\mathrm{d}f) = \mathrm{d}f$. Next we claim that $\mathscr{L}_f$ is continuous for every $f \in \mathrm{C}^\infty(\mathsf{M})$ if $\Gamma^\infty(\mathsf{TM})$ is provided with CO^∞-topology. Indeed, let $K \subseteq \mathsf{M}$, let $m \in \mathbb{Z}_{>0}$, and let $f \in \mathrm{C}^\infty(\mathsf{M})$. By (3.3) (after a few moments of thought), we have, for some suitable $M_0, M_1 \dots, M_m \in \mathbb{R}_{>0}$,

$$p^\infty_{K,m}(\mathscr{L}_X f) \le \sum_{j=0}^{m} M_{m-j} p^\infty_{K,m-j}(X) p^\infty_{K,j+1}(f) \le \sum_{j=0}^{m} M'_j p^\infty_{K,j}(X).$$

This gives continuity of the identity map, if we provide the domain with the CO^∞-topology and the codomain with the weak-$\mathscr{L}$ topology, cf. [22, §III.1.1]. Thus open sets in the weak-$\mathscr{L}$ topology are contained in the CO^∞-topology. □

With respect to the concepts of interest to us, this gives the following result.

Corollary 3.6 (Weak-$\mathscr{L}$ characterisations of Boundedness, Continuity, Measurability, and Integrability for the CO^∞-Topology). *Let M be a smooth manifold, let $(\mathcal{X}, \mathscr{O})$ be a topological space, let $(\mathcal{T}, \mathscr{M})$ be a measurable space, and let $\mu\colon \mathscr{M} \to \overline{\mathbb{R}}_{\ge 0}$ be a finite measure. The following statements hold:*

(i) a subset $\mathcal{B} \subseteq \Gamma^\infty(\mathsf{TM})$ is bounded in the von Neumann bornology if and only if it is weak-$\mathscr{L}$ bounded in the von Neumann bornology;
(ii) a map $\Phi\colon \mathcal{X} \to \Gamma^\infty(\mathsf{TM})$ is continuous if and only if it is weak-$\mathscr{L}$ continuous;
(iii) a map $\Psi\colon \mathcal{T} \to \Gamma^\infty(\mathsf{TM})$ is measurable if and only if it is weak-$\mathscr{L}$ measurable;
(iv) a map $\Psi\colon \mathcal{T} \to \Gamma^\infty(\mathsf{TM})$ is Bochner integrable if and only if it is weak-$\mathscr{L}$ Bochner integrable.

Proof: We first claim that $\mathscr{A} \triangleq \{\mathscr{L}_f \mid f \in \mathrm{C}^\infty(\mathsf{M})\}$ has a countable point separating subset. This is easily proved as follows. For notational simplicity, suppose that M has a well-defined dimension. Let $x \in \mathsf{M}$ and note that there exist a neighbourhood $\mathcal{U}_x$ of x and $f^1_x, \dots, f^n_x \in \mathrm{C}^\infty(\mathsf{M})$ such that

$$\mathsf{T}^*_y\mathsf{M} = \mathrm{span}_{\mathbb{R}}(\mathrm{d}f^1_x(y), \dots, \mathrm{d}f^n_x(y)), \qquad y \in \mathcal{U}_x.$$

Since M is second countable it is Lindelöf [28, Theorem 16.9]. Therefore, there exists $(x_j)_{j\in\mathbb{Z}_{>0}}$ such that $\mathsf{M} = \cup_{j\in\mathbb{Z}_{>0}} \mathcal{U}_{x_j}$. The countable collection of linear mappings

$\mathscr{L}_{f_{x_j}^k}$, $k \in \{1, \dots, n\}$, $j \in \mathbb{Z}_{>0}$, is then point separating. Indeed, if $X, Y \in \Gamma^\infty(\mathsf{TM})$ are distinct, then there exists $x \in \mathsf{M}$ such that $X(x) \neq Y(x)$. Let $j \in \mathbb{Z}_{>0}$ be such that $x \in \mathcal{U}_{x_j}$ and note that we must have $\mathscr{L}_{f_{x_j}^k}(X)(x) \neq \mathscr{L}_{f_{x_j}^k}(Y)(x)$ for some $k \in \{1, \dots, n\}$, giving our claim.

The result is now a direct consequence of Lemma 3.3, noting that the CO^∞-topology on $\Gamma^\infty(\mathsf{TM})$ is complete, separable, and Suslin (we also need that the CO^∞-topology on $\mathrm{C}^\infty(\mathsf{M})$ is Suslin, which it is), as we have seen above in properties CO^∞-2, CO^∞-4, and CO^∞-6. □

3.4 Topologies for Finitely Differentiable Vector Fields

The constructions of this section so far are easily adapted to the case where objects are only finitely differentiable. We sketch here how this can be done. We let $\pi \colon \mathsf{E} \to \mathsf{M}$ be a smooth vector bundle, and we suppose that we have a linear connection ∇^0 on E, an affine connection ∇ on M, a fibre metric $\mathbb{G}_0$ on E, and a Riemannian metric $\mathbb{G}$ on M. Let $r \in \mathbb{Z}_{\geq 0} \cup \{\infty\}$ and let $m \in \mathbb{Z}_{\geq 0}$ with $m \leq r$. By $\Gamma^r(\mathsf{E})$ we denote the space of C^r-sections of E. We define seminorms p_K^m, $K \subseteq \mathsf{M}$ compact, on $\Gamma^r(\mathsf{E})$ by

$$p_K^m(\xi) = \sup\{\|j_m\xi(x)\|_{\overline{\mathbb{G}}_m} \mid x \in K\},$$

and these seminorms define a locally convex topology that we call the ***CO^m-topology***. Let us list some of the attributes of this topology.

CO^m-1. It is Hausdorff: [16, 4.3.1].

CO^m-2. It is complete if and only if $m = r$: [16, 4.3.2].

CO^m-3. It is metrisable: [16, 4.3.1].

CO^m-4. It is separable: This can be shown to follow by an argument similar to that given above for the CO^∞-topology.

CO^m-5. It is probably not nuclear: In case M is compact, note that p_{M}^m is a norm that characterises the CO^m-topology. A normed vector space is nuclear if and only if it is finite-dimensional [18, Theorem 4.4.14], so the CO^m-topology cannot be nuclear when M is compact except in cases of degenerate dimension. But, even when M is not compact, the CO^m-topology is not likely nuclear, although we have neither found a reference nor proved this.

CO^m-6. It is Suslin when $m = r$: This follows since $\Gamma^m(\mathsf{TM})$ is a Polish space, as we have already seen.

CO^m-7. The CO^m-topology is weaker than the CO^r-topology: This is more or less clear from the definitions.

From the preceding, we point out two places where one must take care in using the CO^m-topology, $m \in \mathbb{Z}_{\geq 0}$, contrasted with the CO^∞-topology. First of all, the topology, if used on $\Gamma^r(\mathsf{E})$, $r > m$, is not complete, so convergence arguments must be modified appropriately. Second, it is no longer necessarily the case that bounded sets are relatively compact. Instead, relatively compact subsets will be described by an appropriate version of the Arzelà–Ascoli Theorem, cf. [13, Theorem 5.21].

Therefore, we need to specify for these spaces whether we will be using the von Neumann bornology or the compact bornology when we use the word "bounded". These caveats notwithstanding, it is oftentimes appropriate to use these weaker topologies.

Of course, the preceding can be specialised to vector fields and functions, and one can define the weak-$\mathscr{L}$ topologies corresponding to the topologies for finitely differentiable sections. In doing this, we apply the general construction of Definition 3.2 with $\mathsf{U} = \Gamma^r(\mathsf{TM})$, $\mathsf{V} = \mathrm{C}^r(\mathsf{M})$ (with the CO^m-topology), and $\mathscr{A} = \{\mathscr{L}_f \mid f \in \mathrm{C}^\infty(\mathsf{M})\}$, where

$$\begin{aligned} \mathscr{L}_f \colon \Gamma^r(\mathsf{TM}) &\to \mathrm{C}^r(\mathsf{M}) \\ X &\mapsto \mathscr{L}_X f. \end{aligned}$$

This gives the following definition.

Definition 3.7 (Weak-$\mathscr{L}$ Topology for Space of Finitely Differentiable Vector Fields). Let M be a smooth manifold, let $m \in \mathbb{Z}_{\geq 0}$, and let $r \in \mathbb{Z}_{\geq 0} \cup \{\infty\}$ have the property that $r \geq m$. The ***weak-($\mathscr{L}, m$) topology*** for $\Gamma^r(\mathsf{TM})$ is the weakest topology for which $\mathscr{L}_f$ is continuous for each $f \in \mathrm{C}^\infty(\mathsf{M})$, where $\mathrm{C}^r(\mathsf{M})$ is given the CO^m-topology. ○

We can show that the weak-($\mathscr{L}, m$) topology agrees with the CO^m-topology.

Theorem 3.8 (Weak-$\mathscr{L}$ Topology for Finitely Differentiable Vector Fields). *Let* M *be a smooth manifold, let* $m \in \mathbb{Z}_{\geq 0}$, *and let* $r \in \mathbb{Z}_{\geq 0} \cup \{\infty\}$ *have the property that* $r \geq m$. *Then the following two topologies for* $\Gamma^r(\mathsf{TM})$ *agree:*
(i) the CO^m*-topology;*
(ii) the weak-($\mathscr{L}, m$*)-topology.*

Proof: Let us first show that the CO^m-topology is weaker than the weak-($\mathscr{L}, m$) topology. Just as in the corresponding part of the proof of Theorem 3.5, we can show that, for $K \subseteq \mathsf{M}$ compact, there exist $f^1, \dots, f^r \in \mathrm{C}^\infty(\mathsf{M})$, compact $K_1, \dots, K_r \subseteq \mathsf{M}$, and $C_1, \dots, C_r \in \mathbb{R}_{>0}$ such that

$$p_K^m(X) \leq C_1 p_{K_1}^m(\mathscr{L}_X f^1) + \cdots + C_r p_{K_r}^m(\mathscr{L}_X f^r)$$

for every $X \in \Gamma^r(\mathsf{TM})$. This estimate gives this part of the theorem.

To prove that the weak ($\mathscr{L}, m$)-topology is weaker than the CO^m-topology, it suffices to show that $\mathscr{L}_f$ is continuous if $\Gamma^r(\mathsf{TM})$ and $\mathrm{C}^r(\mathsf{M})$ are given the CO^m-topology. This can be done just as in Theorem 3.5, with suitable modifications since we only have to account for m derivatives. □

We also have the corresponding relationships between various attributes and their weak counterparts.

Corollary 3.9 (Weak-$\mathscr{L}$ Characterisations of Boundedness, Continuity, Measurability, and Integrability for the CO^m-Topology). *Let* M *be a smooth manifold, let* $m \in \mathbb{Z}_{\geq 0}$, *and let* $r \in \mathbb{Z}_{\geq 0} \cup \{\infty\}$ *have the property that* $r \geq m$. *Let* $(\mathcal{X}, \mathscr{O})$ *be a topological space, let* $(\mathcal{T}, \mathscr{M})$ *be a measurable space, and let* $\mu \colon \mathscr{M} \to \overline{\mathbb{R}}_{\geq 0}$ *be a finite measure. The following statements hold:*

(i) *a subset* $\mathcal{B} \subseteq \Gamma^r(\mathsf{TM})$ *is* CO^m*-bounded in the von Neumann bornology if and only if it is weak-*$(\mathscr{L}, m)$ *bounded in the von Neumann bornology;*
(ii) *a map* $\Phi\colon \mathcal{X} \to \Gamma^r(\mathsf{TM})$ *is* CO^m*-continuous if and only if it is weak-*$(\mathscr{L}, m)$ *continuous;*
(iii) *a map* $\Psi\colon \mathcal{T} \to \Gamma^m(\mathsf{TM})$ *is* CO^m*-measurable if and only if it is weak-*$(\mathscr{L}, m)$ *measurable;*
(iv) *a map* $\Psi\colon \mathcal{T} \to \Gamma^m(\mathsf{TM})$ *is Bochner integrable if and only if it is weak-*$(\mathscr{L}, m)$ *Bochner integrable.*

Proof: In the proof of Corollary 3.6 we established that $\{\mathscr{L}_f \mid f \in \mathrm{C}^\infty(\mathsf{M})\}$ was point separating as a family of linear mappings with domain $\Gamma^\infty(\mathsf{TM})$. The same proof is valid if the domain is $\Gamma^m(\mathsf{TM})$. The result is then a direct consequence of Lemma 3.3, taking care to note that the CO^m-topology on $\Gamma^r(\mathsf{TM})$ is separable, and is also complete and Suslin when $r = m$ (and $\mathrm{C}^r(\mathsf{M})$ is Suslin when $r = m$), as we have seen in properties CO^m-2, CO^m-4, and CO^m-6 above. □

3.5 Topologies for Lipschitz Vector Fields

It is also possible to characterise Lipschitz sections, so let us indicate how this is done in geometric terms. Throughout our discussion of the Lipschitz case, we make the assumption that the affine connection ∇ on M is the Levi–Civita connection for $\mathbb{G}$ and that the linear connection ∇^0 on E is $\mathbb{G}_0$-orthogonal, by which we mean that parallel translation consists of isometries. The existence of such a connection is ensured by the reasoning of Kobayashi and Nomizu [14] following the proof of their Proposition III.1.5. We suppose that M is connected, for simplicity. If it is not, then one has to allow the metric we are about to define to take infinite values. This is not problematic [5, Exercise 1.1.2], but we wish to avoid the more complicated accounting procedures. The ***length*** of a piecewise differentiable curve $\gamma\colon [a, b] \to \mathsf{M}$ is

$$\ell_{\mathbb{G}}(\gamma) = \int_a^b \sqrt{\mathbb{G}(\gamma'(t), \gamma'(t))}\, \mathrm{d}t.$$

One easily shows that the length of the curve γ depends only on image(γ) and not on the particular parameterisation. We can, therefore, restrict ourselves to curves defined on $[0, 1]$. In this case, for $x_1, x_2 \in \mathsf{M}$, we define the ***distance*** between x_1 and x_2 to be

$$\mathrm{d}_{\mathbb{G}}(x_1, x_2) = \inf\{\ell_{\mathbb{G}}(\gamma) \mid \gamma\colon [0, 1] \to \mathsf{M} \text{ is a piecewise differentiable curve for which } \gamma(0) = x_1 \text{ and } \gamma(1) = x_2\}.$$

It is relatively easy to show that $(\mathsf{M}, \mathrm{d}_{\mathbb{G}})$ is a metric space [1, Proposition 5.5.10].

Now we define a canonical Riemannian metric on the total space E of a vector bundle $\pi\colon \mathsf{E} \to \mathsf{M}$, following the construction of Sasaki [21] for tangent bundles. The linear connection ∇^0 gives a splitting $\mathsf{TE} \simeq \pi^*\mathsf{TM} \oplus \pi^*\mathsf{E}$ [15, §11.11]. The second component of this decomposition is the vertical component so $T_{e_x}\pi$ restricted to the first component is an isomorphism onto $\mathsf{T}_x\mathsf{M}$, i.e., the first component is "horizontal". Let us denote by hor: $\mathsf{TE} \to \pi^*\mathsf{TM}$ and ver: $\mathsf{TE} \to \pi^*\mathsf{E}$ the projections onto the first and second components of the direct sum decomposition. This then gives the Riemannian metric $\mathbb{G}_\mathsf{E}$ on E defined by

$$\mathbb{G}_\mathsf{E}(X_{e_x}, Y_{e_x}) = \mathbb{G}(\mathrm{hor}(X_{e_x}), \mathrm{hor}(Y_{e_x})) + \mathbb{G}_0(\mathrm{ver}(X_{e_x}), \mathrm{ver}(Y_{e_x})).$$

Now let us consider various ways of characterising Lipschitz sections. To this end, we let $\xi\colon \mathsf{M} \to \mathsf{E}$ be such that $\xi(x) \in \mathsf{E}_x$ for every $x \in \mathsf{M}$. For compact $K \subseteq \mathsf{M}$ we then define

$$L_K(\xi) = \sup\left\{ \frac{\mathrm{d}_{\mathbb{G}_\mathsf{E}}(\xi(x_1), \xi(x_2))}{\mathrm{d}_\mathbb{G}(x_1, x_2)} \;\middle|\; x_1, x_2 \in K,\ x_1 \neq x_2 \right\}.$$

This is the ***K*-dilatation** of ξ. For a piecewise differentiable curve $\gamma\colon [0, T] \to \mathsf{M}$, we denote by $\tau_{\gamma,t}\colon \mathsf{E}_{\gamma(0)} \to \mathsf{E}_{\gamma(t)}$ the isomorphism of parallel translation along γ for each $t \in [0, T]$. We then define

$$l_K(\xi) = \sup\Bigl\{ \frac{\|\tau_{\gamma,1}^{-1}(\xi\circ\gamma(1)) - \xi\circ\gamma(0)\|_{\mathbb{G}_0}}{\ell_\mathbb{G}(\gamma)} \;\Big|\; \gamma\colon [0,1] \to \mathsf{M},\ \gamma(0), \gamma(1) \in K,\ \gamma(0) \neq \gamma(1) \Bigr\}, \tag{3.4}$$

which is the ***K*-sectional dilatation** of ξ. Finally, we define

$$\begin{aligned} \mathrm{Dil}\,\xi\colon \mathsf{M} &\to \mathbb{R}_{\geq 0} \\ x &\mapsto \inf\{L_{\mathrm{cl}(\mathcal{U})}(\xi) \mid \mathcal{U} \text{ is a relatively compact neighbourhood of } x\}, \end{aligned}$$

and

$$\begin{aligned} \mathrm{dil}\,\xi\colon \mathsf{M} &\to \mathbb{R}_{\geq 0} \\ x &\mapsto \inf\{l_{\mathrm{cl}(\mathcal{U})}(\xi) \mid \mathcal{U} \text{ is a relatively compact neighbourhood of } x\}, \end{aligned}$$

which are the ***local dilatation*** and ***local sectional dilatation***, respectively, of ξ. Following [27, Proposition 1.5.2] one can show that

$$L_K(\xi + \eta) \leq L_K(\xi) + L_K(\eta), \quad l_K(\xi + \eta) \leq l_K(\xi) + l_K(\eta), \qquad K \subseteq \mathsf{M} \text{ compact},$$

and

$$\mathrm{Dil}\,(\xi + \eta)(x) \leq \mathrm{Dil}\,\xi(x) + \mathrm{Dil}\,\eta(x), \quad \mathrm{dil}\,(\xi + \eta)(x) \leq \mathrm{dil}\,\xi(x) + \mathrm{dil}\,\eta(x), \qquad x \in \mathsf{M}.$$

The following lemma connects the preceding notions.

Lemma 3.10 (Characterisations of Lipschitz Sections). *Let $\pi\colon \mathsf{E} \to \mathsf{M}$ be a smooth vector bundle and let $\xi\colon \mathsf{M} \to \mathsf{E}$ be such that $\xi(x) \in \mathsf{E}_x$ for every $x \in \mathsf{M}$. Then the following statements are equivalent:*

(i) $L_K(\xi) < \infty$ for every compact $K \subseteq \mathsf{M}$;
(ii) $l_K(\xi) < \infty$ for every compact $K \subseteq \mathsf{M}$;
(iii) $\mathrm{Dil}\,\xi(x) < \infty$ for every $x \in \mathsf{M}$;
(iv) $\mathrm{dil}\,\xi(x) < \infty$ for every $x \in \mathsf{M}$.

Moreover, we have the equalities

$$L_K(\xi) = \sqrt{l_K(\xi)^2 + 1}, \quad \mathrm{Dil}\,\xi(x) = \sqrt{\mathrm{dil}\,\xi(x)^2 + 1}$$

for every compact $K \subseteq \mathsf{M}$ and every $x \in \mathsf{M}$.

Proof: The equivalence of (i) and (ii), along with the equality $L_K = \sqrt{l_K^2 + 1}$, follows from the arguments of [6, Lemma II.A.2.4]. This also implies the equality $\mathrm{Dil}\,\xi(x) = \sqrt{\mathrm{dil}\,\xi(x)^2 + 1}$ when both $\mathrm{Dil}\,\xi(x)$ and $\mathrm{dil}\,\xi(x)$ are finite.

(i) $\Longrightarrow$ (iii) If $x \in \mathsf{M}$ and if $\mathcal{U}$ is a relatively compact neighbourhood of x, then $L_{\mathrm{cl}(\mathcal{U})}(\xi) < \infty$ and so $\mathrm{Dil}\,\xi(x) < \infty$.

(ii) $\Longrightarrow$ (iv) This follows just as does the preceding part of the proof.

(iii) $\Longrightarrow$ (i) Suppose that $\mathrm{Dil}\,\xi(x) < \infty$ for every $x \in \mathsf{M}$ and that there exists a compact set $K \subseteq \mathsf{M}$ such that $L_K(\xi) \not< \infty$. Then there exist sequences $(x_j)_{j\in\mathbb{Z}_{>0}}$ and $(y_j)_{j\in\mathbb{Z}_{>0}}$ in K such that $x_j \neq y_j$, $j \in \mathbb{Z}_{>0}$, and

$$\lim_{j\to\infty} \frac{\mathrm{d}_{\mathbb{G}_\mathsf{E}}(\xi(x_j), \xi(y_j))}{\mathrm{d}_{\mathbb{G}}(x_j, y_j)} = \infty.$$

Since $\mathrm{Dil}\,\xi(x) < \infty$ for every $x \in \mathsf{M}$, it follows directly that ξ is continuous and so $\xi(K)$ is bounded in the metric $\mathbb{G}_\mathsf{E}$. Therefore, there exists $C \in \mathbb{R}_{>0}$ such that

$$\mathrm{d}_{\mathbb{G}_\mathsf{E}}(\xi(x_j), \xi(y_j)) \leq C, \qquad j \in \mathbb{Z}_{>0},$$

and so we must have $\lim_{j\to\infty} \mathrm{d}_{\mathbb{G}}(x_j, y_j) = 0$. Let $(x_{j_k})_{k\in\mathbb{Z}_{>0}}$ be a subsequence converging to $x \in K$ and note that $(y_{j_k})_{k\in\mathbb{Z}_{>0}}$ then also converges to x. This implies that $\mathrm{Dil}\,\xi(x) \not< \infty$, which proves the result.

(iv) $\Longrightarrow$ (ii) This follows just as the preceding part of the proof. □

With the preceding, we can define what we mean by a locally Lipschitz section of a vector bundle, noting that, if $\mathrm{dil}\,\xi(x) < \infty$ for every $x \in \mathsf{M}$, ξ is continuous. Our definition is in the general situation where sections are of class C^m with the mth derivative being, not just continuous, but Lipschitz.

Definition 3.11 (Locally Lipschitz Section). For a smooth vector bundle $\pi\colon \mathsf{E} \to \mathsf{M}$ and for $m \in \mathbb{Z}_{\geq 0}$, $\xi \in \Gamma^m(\mathsf{E})$ is of ***class $\mathrm{C}^{m+\mathrm{lip}}$*** if $j_m\xi\colon \mathsf{M} \to \mathsf{J}^m\mathsf{E}$ satisfies any of the four equivalent conditions of Lemma 3.10. If ξ is of class $\mathrm{C}^{0+\mathrm{lip}}$, then we say it is ***locally Lipschitz***. By $\Gamma^{\mathrm{lip}}(\mathsf{E})$ we denote the space of locally Lipschitz sections of E. For $m \in \mathbb{Z}_{\geq 0}$, by $\Gamma^{m+\mathrm{lip}}(\mathsf{E})$ we denote the space of sections of E of class $\mathrm{C}^{m+\mathrm{lip}}$. ○

It is straightforward, if tedious, to show that a section is of class $C^{m+\mathrm{lip}}$ if and only if, in any coordinate chart, the section is m-times continuously differentiable with the mth derivative being locally Lipschitz in the usual Euclidean sense. The essence of the argument is that, in any sufficiently small neighbourhood of a point in M, the distance functions $d_{\mathbb{G}}$ and $d_{\mathbb{G}_{\mathsf{E}}}$ are equivalent to the Euclidean distance functions defined in coordinates.

The following characterisation of the local sectional dilatation is useful.

Lemma 3.12 (Local Sectional Dilatation Using Derivatives). *For a smooth vector bundle $\pi\colon \mathsf{E} \to \mathsf{M}$ and for $\xi \in \Gamma^{\mathrm{lip}}(\mathsf{E})$, we have*

$$\operatorname{dil}\xi(x) = \inf\{\sup\{\|\nabla^0_{v_y}\xi\|_{\mathbb{G}_0} \mid y \in \mathrm{cl}(\mathcal{U}),\ \|v_y\|_{\mathbb{G}} = 1,\ \xi \textit{ differentiable at } y\} \mid \mathcal{U} \textit{ is a relatively compact neighbourhood of } x\}.$$

Proof: As per [14, Proposition IV.3.4], let $\mathcal{U}$ be a geodesically convex, relatively compact open set. We claim that

$$l_{\mathrm{cl}(\mathcal{U})}(\xi) = \sup\{\|\nabla^0_{v_y}\xi\|_{\mathbb{G}_0} \mid y \in \mathrm{cl}(\mathcal{U}),\ \|v_y\|_{\mathbb{G}} = 1,\ \xi \text{ differentiable at } y\}.$$

By [6, Lemma II.A.2.4], to determine $l_{\mathrm{cl}(\mathcal{U})}(\xi)$, it suffices in the formula (3.4) to use only length minimising geodesics whose images are contained in $\mathrm{cl}(\mathcal{U})$. Let $x \in \mathcal{U}$, let $v_x \in \mathsf{T}_x\mathsf{M}$ have unit length, and let $\gamma\colon [0,T] \to \mathrm{cl}(\mathcal{U})$ be a minimal length geodesic such that $\gamma'(0) = v_x$. If x is a point of differentiability for ξ, then

$$\lim_{t\to 0} \frac{\|\tau^{-1}_{\gamma,t}(\xi\circ\gamma(t)) - \xi\circ\gamma(0)\|_{\mathbb{G}_0}}{t} = \|\nabla^0_{v_y}\xi\|_{\mathbb{G}_0}.$$

From this we conclude that

$$l_{\mathrm{cl}(\mathcal{U})}(\xi) \geq \sup\{\|\nabla^0_{v_x}\xi\|_{\mathbb{G}_0} \mid x \in \mathrm{cl}(\mathcal{U}),\ \|v_x\|_{\mathbb{G}} = 1,\ \xi \text{ differentiable at } y\}.$$

Suppose the opposite inequality does not hold. Then there exist $x_1, x_2 \in \mathrm{cl}(\mathcal{U})$ such that, if $\gamma\colon [0,T] \to \mathsf{M}$ is the arc-length parameterised minimal length geodesic from x_1 to x_2, then

$$\frac{\|\tau^{-1}_{\gamma,T}(\xi\circ\gamma(T)) - \xi\circ\gamma(0)\|}{T} > \|\nabla^0_{v_x}\xi\|_{\mathbb{G}_0} \tag{3.5}$$

for every $x \in \mathrm{cl}(\mathcal{U})$ for which ξ is differentiable at x and every $v_x \in \mathsf{T}_x\mathsf{M}$ of unit length. Note that $\alpha\colon t \mapsto \tau^{-1}_{\gamma,t}(\xi\circ\gamma(t))$ is a Lipschitz curve in $\mathsf{T}_{x_1}\mathsf{M}$. By Rademacher's Theorem [8, Theorem 3.1.5], this curve is almost everywhere differentiable. If α is differentiable at t we have

$$\alpha'(t) = \tau^{-1}_{\gamma,t}(\nabla^0_{\gamma'(t)}\xi).$$

Therefore, also by Rademacher's Theorem and since ∇^0 is $\mathbb{G}_0$-orthogonal, we have

$$\sup\left\{ \frac{\|\tau_{\gamma,t}^{-1}(\xi\circ\gamma(t)) - \xi\circ\gamma(0)\|_{\mathbb{G}_0}}{t} \;\middle|\; t\in[0,T]\right\} = \sup\{\|\nabla^0_{\gamma'(t)}\xi\|_{\mathbb{G}_0} \mid t\in[0,T],\ \xi \text{ is differentiable at } \gamma(t)\}.$$

This, however, contradicts (3.5), and so our claim holds.

Now let $x \in \mathsf{M}$ and let $(\mathcal{U}_j)_{j\in\mathbb{Z}_{>0}}$ be a sequence of relatively compact, geodesically convex neighbourhood of x such that $\cap_{j\in\mathbb{Z}_{>0}}\mathcal{U}_j = \{x\}$. Then

$$\operatorname{dil}\xi(x) = \lim_{j\to\infty} l_{\operatorname{cl}(\mathcal{U}_j)}(\xi)$$

and

$$\inf\{\sup\{\|\nabla^0_{v_y}\xi\|_{\mathbb{G}_0} \mid y\in\operatorname{cl}(\mathcal{U}),\ \|v_y\|_{\mathbb{G}} = 1,\ \xi \text{ differentiable at } y\} \mid \mathcal{U} \text{ is a relatively compact neighbourhood of } x\} = \lim_{j\to\infty}\sup\{\|\nabla^0_{v_y}\xi\|_{\mathbb{G}_0} \mid y\in\operatorname{cl}(\mathcal{U}_j),\ \|v_y\|_{\mathbb{G}} = 1,\ \xi \text{ differentiable at } y\}.$$

The lemma now follows from the claim in the opening paragraph. □

Let us see how to topologise spaces of locally Lipschitz sections. Lemma 3.10 gives us four possibilities for doing this. In order to be as consistent as possible with our other definitions of seminorms, we use the "locally sectional" characterisation of Lipschitz seminorms. Thus, for $\xi \in \Gamma^{\mathrm{lip}}(\mathsf{E})$ and $K \subseteq \mathsf{M}$ compact, let us define

$$\lambda_K(\xi) = \sup\{\operatorname{dil}\xi(x) \mid x\in K\}$$

and then define a seminorm p_K^{lip}, $K \subseteq \mathsf{M}$ compact, on $\Gamma^{\mathrm{lip}}(\mathsf{E})$ by

$$p_K^{\mathrm{lip}}(\xi) = \max\{\lambda_K(\xi), p_K^0(\xi)\}.$$

The seminorms p_K^{lip}, $K \subseteq \mathsf{M}$ compact, give the **CO^{lip}*-topology*** on $\Gamma^r(\mathsf{E})$ for $r \in \mathbb{Z}_{>0}\cup\{\infty\}$. To topologise $\Gamma^{m+\mathrm{lip}}(\mathsf{E})$, note that the CO^{lip}-topology on $\Gamma^{\mathrm{lip}}(\mathsf{J}^m\mathsf{E})$ induces a topology on $\Gamma^{m+\mathrm{lip}}(\mathsf{E})$ that we call the **CO$^{m+\mathrm{lip}}$*-topology***. The seminorms for this locally convex topology are

$$p_K^{m+\mathrm{lip}}(\xi) = \max\{\lambda_K^m(\xi), p_K^m(\xi)\}, \qquad K\subseteq \mathsf{M} \text{ compact},$$

where

$$\lambda_K^m(\xi) = \sup\{\operatorname{dil} j_m\xi(x) \mid x\in K\}.$$

Note that $\operatorname{dil} j_m\xi$ is unambiguously defined. Let us briefly explain why. If the connections ∇ and ∇^0 are metric connections for $\mathbb{G}$ and $\mathbb{G}_0$, as we are assuming, then the induced connection ∇^m on $\mathsf{T}^k(\mathsf{T}^*\mathsf{M})\otimes\mathsf{E}$ is also metric with respect to the induced metric determined from Lemma 2.3. It then follows from Lemma 2.1 that the dilatation for sections of $\mathsf{J}^m\mathsf{E}$ can be defined just as for sections of E.

Note that $\Gamma^{\mathrm{lip}}(\mathsf{E}) \subseteq \Gamma^0(\mathsf{E})$ and $\Gamma^r(\mathsf{E}) \subseteq \Gamma^{\mathrm{lip}}(\mathsf{E})$ for $r \in \mathbb{Z}_{>0}$. Thus we adopt the convention that $0 < \mathrm{lip} < 1$ for the purposes of ordering degrees of regularity. Let $m \in \mathbb{Z}_{\geq 0}$, and let $r \in \mathbb{Z}_{\geq 0} \cup \{\infty\}$ and $r' \in \{0, \mathrm{lip}\}$ be such that $r + r' \geq m + \mathrm{lip}$. We adopt the obvious convention that $\infty + \mathrm{lip} = \infty$. The seminorms $p_K^{m+\mathrm{lip}}$, $K \subseteq \mathsf{M}$ compact, can then be defined on $\Gamma^{r+r'}(\mathsf{E})$.

Let us record some properties of the $\mathrm{CO}^{m+\mathrm{lip}}$-topology for $\Gamma^{r+r'}(\mathsf{E})$. This topology is not extensively studied like the other differentiable topologies, but we can nonetheless enumerate its essential properties.

$\mathrm{CO}^{m+\mathrm{lip}}$-1. It is Hausdorff: This is clear.

$\mathrm{CO}^{m+\mathrm{lip}}$-2. It is complete if and only if $r + r' = m + \mathrm{lip}$: This is more or less because, for a compact metric space, the space of Lipschitz functions is a Banach space [27, Proposition 1.5.2]. Since $\Gamma^{m+\mathrm{lip}}(\mathsf{E})$ is the inverse limit of the Banach spaces $\Gamma^{m+\mathrm{lip}}(\mathsf{E}|K_j)$,[2] $j \in \mathbb{Z}_{>0}$, for a compact exhaustion $(K_j)_{j\in\mathbb{Z}_{>0}}$ of M, and since the inverse limit of complete locally convex spaces is complete [11, Proposition 2.11.3], we conclude the stated assertion.

$\mathrm{CO}^{m+\mathrm{lip}}$-3. It is metrisable: This is argued as follows. First of all, it is a countable inverse limit of Banach spaces. Inverse limits are closed subspaces of the direct product [19, Proposition V.19]. The direct product of metrisable spaces, in particular Banach spaces, is metrisable [28, Theorem 22.3].

$\mathrm{CO}^{m+\mathrm{lip}}$-4. It is separable: This is a consequence of the result of [9, Theorem 1.2′] which says that Lipschitz functions on Riemannian manifolds can be approximated in the $\mathrm{CO}^{\mathrm{lip}}$-topology by smooth functions, and by the separability of the space of smooth functions.

$\mathrm{CO}^{m+\mathrm{lip}}$-5. It is probably not nuclear: For compact base manifolds, $\Gamma^{m+\mathrm{lip}}(\mathsf{E})$ is an infinite-dimensional normed space, and so not nuclear [18, Theorem 4.4.14]. But, even when M is not compact, the $\mathrm{CO}^{m+\mathrm{lip}}$-topology is not likely nuclear, although we have neither found a reference nor proved this.

$\mathrm{CO}^{m+\mathrm{lip}}$-6. It is Suslin when $m + \mathrm{lip} = r + r'$: This follows since $\Gamma^{m+\mathrm{lip}}(\mathsf{E})$ is a Polish space, as we have already seen.

Of course, the preceding can be specialised to vector fields and functions, and one can define the weak-$\mathscr{L}$ topologies corresponding to the above topologies. To do this, we apply the general construction of Definition 3.2 with $\mathsf{U} = \Gamma^{r+r'}(\mathsf{TM})$, $\mathsf{V} = \mathrm{C}^{r+r'}(\mathsf{M})$ (with the $\mathrm{CO}^{m+\mathrm{lip}}$-topology), and $\mathscr{A} = \{\mathscr{L}_f \mid f \in \mathrm{C}^\infty(\mathsf{M})\}$, where

$$\begin{aligned} \mathscr{L}_f\colon \Gamma^{r+r'}(\mathsf{TM}) &\to \mathrm{C}^{r+r'}(\mathsf{M}) \\ X &\mapsto \mathscr{L}_X f. \end{aligned}$$

We then have the following definition.

Definition 3.13 (Weak-$\mathscr{L}$ Topology for Space of Lipschitz Vector Fields). Let M be a smooth manifold, let $m \in \mathbb{Z}_{\geq 0}$, and let $r \in \mathbb{Z}_{\geq 0} \cup \{\infty\}$ and $r' \in \{0, \mathrm{lip}\}$ have the property that $r + r' \geq m + \mathrm{lip}$. The ***weak-($\mathscr{L}$, m + lip) topology*** for $\Gamma^{r+r'}(\mathsf{TM})$ is the

[2] To be clear, by $\Gamma^{m+\mathrm{lip}}(\mathsf{E}|K)$ we denote the space of sections of class $m + \mathrm{lip}$ defined on a neighbourhood of K.

weakest topology for which $\mathscr{L}_f$ is continuous for each $f \in \mathrm{C}^\infty(\mathsf{M})$, where $\mathrm{C}^{r+r'}(\mathsf{M})$ is given the $\mathrm{CO}^{m+\mathrm{lip}}$-topology. ∘

We can show that the weak-($\mathscr{L}, m + \mathrm{lip}$) topology agrees with the $\mathrm{CO}^{m+\mathrm{lip}}$-topology.

Theorem 3.14 (Weak-$\mathscr{L}$ Topology for Lipschitz Vector Fields). *Let* M *be a smooth manifold, let* $m \in \mathbb{Z}_{\geq 0}$, *and let* $r \in \mathbb{Z}_{\geq 0} \cup \{\infty\}$ *and* $r' \in \{0, \mathrm{lip}\}$ *have the property that* $r + r' \geq m + \mathrm{lip}$. *Then the following two topologies for* $\Gamma^{r+r'}(\mathsf{E})$ *agree:*
(i) the $\mathrm{CO}^{m+\mathrm{lip}}$*-topology;*
(ii) the weak-$(\mathscr{L}, m + \mathrm{lip})$*-topology.*

Proof: We prove the theorem only for the case $m = 0$, since the general case follows from this in combination with Theorem 3.8.

Let us first show that the $\mathrm{CO}^{\mathrm{lip}}$-topology is weaker than the weak-($\mathscr{L}$, lip) topology. Let $K \subseteq \mathsf{M}$ be compact and for $x \in \mathsf{M}$ choose a coordinate chart $(\mathcal{U}_x, \phi_x)$ and functions $f_x^1, \dots, f_x^n \in \mathrm{C}^\infty(\mathsf{M})$ agreeing with the coordinate functions in a neighbourhood of a geodesically convex relatively compact neighbourhood $\mathcal{V}_x$ of x [14, Proposition IV.3.4]. We denote by $\boldsymbol{X}\colon \phi_x(\mathcal{U}_x) \to \mathbb{R}^n$ the local representative of $X \in \Gamma^{\mathrm{lip}}(\mathsf{TM})$. Since $\mathscr{L}_X f_x^j = X^j$ on a neighbourhood of $\mathcal{V}_x$, there exists $C_x \in \mathbb{R}_{>0}$ such that

$$\|\tau_{\gamma,1}^{-1}(X(x_1)) - X(x_2)\|_{\mathbb{G}} \leq C_x \sum_{j=1}^{n} |\mathscr{L}_X f_x^j(x_1) - \mathscr{L}_X f_x^j(x_2)|, \qquad X \in \Gamma^{\mathrm{lip}}(\mathsf{TM}),$$

for every distinct $x_1, x_2 \in \mathrm{cl}(\mathcal{V}_x)$, where γ is the unique minimal length geodesic from x_2 to x_1 (the inequality is a consequence of the fact that the ℓ^1 norm for $\mathbb{R}^n$ is equivalent to any other norm). This gives an inequality

$$\mathrm{dil}\, X(y) \leq C_x(\mathrm{dil}\, \mathscr{L}_X f_x^1(y) + \cdots + \mathrm{dil}\, \mathscr{L}_X f_x^n(y))$$

for every $y \in \mathcal{V}_x$. Now let $x_1, \dots, x_k \in K$ be such that $K \subseteq \cup_{j=1}^k \mathcal{V}_{x_j}$. From this point, it is a bookkeeping exercise, exactly like that in the corresponding part of the proof of Theorem 3.5, to arrive at the inequality

$$\lambda_K(X) \leq C_1 \lambda_K(\mathscr{L}_X f^1) + \cdots + C_r \lambda_K(\mathscr{L}_X f^r),$$

for suitable $f^1, \dots, f^r$, with $r = kn$. From the proof of Theorem 3.8 we also have

$$p_K^0(X) \leq C_1' p_K^0(\mathscr{L}_X f^1) + \cdots + C_r' p_K^0(\mathscr{L}_X f^r),$$

and this gives the result.

To prove that the weak ($\mathscr{L}$, lip)-topology is weaker than the $\mathrm{CO}^{\mathrm{lip}}$-topology, it suffices to show that $\mathscr{L}_f$ is continuous for every $f \in \mathrm{C}^\infty(\mathsf{M})$ if $\Gamma^{r+r'}(\mathsf{TM})$ and $\mathrm{C}^{r+r'}(\mathsf{M})$ are given the $\mathrm{CO}^{m+\mathrm{lip}}$-topology. Thus let $K \subseteq \mathsf{M}$ be compact and let $f \in \mathrm{C}^\infty(\mathsf{M})$. We choose a relatively compact geodesically convex chart $(\mathcal{U}_x, \phi_x)$ about $x \in K$ and compute, for distinct $x_1, x_2 \in \mathcal{U}_x$,

$$\begin{aligned}
&|\mathscr{L}_X f(x_1) - \mathscr{L}_X f(x_2)| \\
&\leq \sum_{j=1}^{n} \left| X^j(x_1)\frac{\partial f}{\partial x^j}(x_1) - X^j(x_2)\frac{\partial f}{\partial x^j}(x_2)\right| \\
&\leq \sum_{j=1}^{n} \left(|X^j(x_1)| \left|\frac{\partial f}{\partial x^j}(x_1) - \frac{\partial f}{\partial x^j}(x_2)\right| + |X^j(x_1) - X^j(x_2)| \left|\frac{\partial f}{\partial x^j}(x_2)\right| \right) \\
&\leq \sum_{j=1}^{n} \left(A_x p^0_{\mathrm{cl}(\mathcal{U}_x)}(X) \frac{\partial f}{\partial x^j}(y) \mathrm{d}_{\mathbb{G}}(x_1, x_2) \right) + B_x \|\tau_{\gamma,1}^{-1} X(x_1) - X(x_2)\|_{\mathbb{G}},
\end{aligned}$$

for some $y \in \mathcal{U}_x$, using the mean value theorem [1, Proposition 2.4.8], and where γ is the unique length minimising geodesic from x_2 to x_1. Thus we have an inequality

$$\lambda_{\mathrm{cl}(\mathcal{U}_x)}(\mathscr{L}_X f) \leq A_x p^0_{\mathrm{cl}(\mathcal{U}_x)}(X) + B_x \lambda_{\mathrm{cl}(\mathcal{U}_x)}(X),$$

for a possibly different A_x. Letting $x_1, \dots, x_k \in K$ be such that $K \subseteq \cup_{j=1}^{k} \mathcal{U}_x$, some more bookkeeping like that in the first part of the proof of Theorem 3.5 gives

$$\lambda_K(\mathscr{L}_X f) \leq \sum_{j=1}^{k} (A_j p^0_{\mathrm{cl}(\mathcal{U}_{x_j})}(X) + B_j \lambda_{\mathrm{cl}(\mathcal{U}_{x_j})}(X))$$

for suitable constants $A_j, B_j \in \mathbb{R}_{>0}$, $j \in \{1, \dots, k\}$. Since, from the proof of Theorem 3.8, we also have

$$p^0_K(\mathscr{L}_X f) \leq \sum_{j=1}^{k} C_j p^0_{\mathrm{cl}(\mathcal{U}_{x_j})}(X)$$

for suitable constants $C_1, \dots, C_k \in \mathbb{R}_{>0}$, the result follows. □

We also have the corresponding relationships between various attributes and their weak counterparts.

Corollary 3.15 (Weak-$\mathscr{L}$ Characterisations of Boundedness, Continuity, Measurability, and Integrability for the $\mathrm{CO}^{m+\mathrm{lip}}$-Topology). *Let* M *be a smooth manifold, let $m \in \mathbb{Z}_{\geq 0}$, and let $r \in \mathbb{Z}_{\geq 0} \cup \{\infty, \mathrm{lip}\}$ and $r' \in \{0, \mathrm{lip}\}$ have the property that $r + r' \geq m + \mathrm{lip}$. Let $(\mathcal{X}, \mathscr{O})$ be a topological space, let $(\mathcal{T}, \mathscr{M})$ be a measurable space, and let $\mu\colon \mathscr{M} \to \overline{\mathbb{R}}_{\geq 0}$ be a finite measure. The following statements hold:*

(i) a subset $\mathcal{B} \subseteq \Gamma^{r+r'}(\mathsf{TM})$ is $\mathrm{CO}^{m+\mathrm{lip}}$-bounded in the von Neumann bornology if and only if it is weak-$(\mathscr{L}, m + \mathrm{lip})$ bounded in the von Neumann bornology;

(ii) a map $\Phi\colon \mathcal{X} \to \Gamma^{r+r'}(\mathsf{TM})$ is $\mathrm{CO}^{m+\mathrm{lip}}$-continuous if and only if it is weak-$(\mathscr{L}, m + \mathrm{lip})$ continuous;

(iii) a map $\Psi\colon \mathcal{T} \to \Gamma^{m+\mathrm{lip}}(\mathsf{TM})$ is $\mathrm{CO}^{m+\mathrm{lip}}$-measurable if and only if it is weak-$(\mathscr{L}, m + \mathrm{lip})$ measurable;

(iv) a map $\Psi\colon \mathcal{T} \to \Gamma^{m+\mathrm{lip}}(\mathsf{TM})$ is Bochner integrable if and only if it is weak-$(\mathscr{L}, m + \mathrm{lip})$ Bochner integrable.

Proof: In the proof of Corollary 3.6 we established that $\{\mathscr{L}_f \mid f \in \mathrm{C}^\infty(\mathsf{M})\}$ was point separating as a family of linear mappings with domain $\Gamma^\infty(\mathsf{TM})$. The same proof is valid if the domain is $\Gamma^{m+\mathrm{lip}}(\mathsf{TM})$. The result is then a direct consequence of Lemma 3.3, noting that the $\mathrm{CO}^{m+\mathrm{lip}}$-topology on $\Gamma^{r+r'}(\mathsf{TM})$ is separable, and is also complete and Suslin when $r + r' = m + \mathrm{lip}$ (and $\mathrm{C}^{r+r'}(\mathsf{M})$ is Suslin when $r + r' = m + \mathrm{lip}$), as we have seen above in properties $\mathrm{CO}^{m+\mathrm{lip}}$-2, $\mathrm{CO}^{m+\mathrm{lip}}$-4, and $\mathrm{CO}^{m+\mathrm{lip}}$-6. □

Notation 3.16 ($m + m'$). In order to try to compactify the presentation of the various degrees of regularity we consider, we will frequently speak of the class "$m + m'$" where $m \in \mathbb{Z}_{\geq 0}$ and $m' \in \{0, \mathrm{lip}\}$. This allows us to include the various Lipschitz cases alongside the finitely differentiable cases. Thus, whenever the reader sees "$m + m'$", this is what they should have in mind. ∘

References

1. Abraham, R., Marsden, J.E., Ratiu, T.S.: Manifolds, Tensor Analysis, and Applications, 2 edn. No. 75 in Applied Mathematical Sciences. Springer-Verlag (1988)
2. Agrachev, A.A., Sachkov, Y.: Control Theory from the Geometric Viewpoint, *Encyclopedia of Mathematical Sciences*, vol. 87. Springer-Verlag, New York/Heidelberg/Berlin (2004)
3. Beckmann, R., Deitmar, A.: Strong vector valued integrals (2011). URL http://arxiv.org/abs/1102.1246v1. ArXiv:1102.1246v1 [math.FA]
4. Bogachev, V.I.: Measure Theory, vol. 2. Springer-Verlag, New York/Heidelberg/Berlin (2007)
5. Burago, D., Burago, Y., Ivanov, S.: A Course in Metric Geometry. No. 33 in Graduate Studies in Mathematics. American Mathematical Society, Providence, RI (2001)
6. Canary, R.D., Epstein, D.B.A., Marden, A. (eds.): Fundamentals of Hyperbolic Geometry: Selected Expositions. No. 328 in London Mathematical Society Lecture Note Series. Cambridge University Press, New York/Port Chester/Melbourne/Sydney (2006)
7. Dodson, C.T.J., Poston, T.: Tensor Geometry. No. 130 in Graduate Texts in Mathematics. Springer-Verlag, New York/Heidelberg/Berlin (1991)
8. Federer, H.: Geometric Measure Theory. Classics in Mathematics. Springer-Verlag, New York/Heidelberg/Berlin (1996). Reprint of 1969 edition
9. Greene, R.E., Wu, H.: C^∞-approximations of convex, subharmonic, and plurisubharmonic functions. Annales Scientifiques de l'École Normale Supérieure. Quatrième Série **12**(1), 47–84 (1979)
10. Hirsch, M.W.: Differential Topology. No. 33 in Graduate Texts in Mathematics. Springer-Verlag, New York/Heidelberg/Berlin (1976)
11. Horváth, J.: Topological Vector Spaces and Distributions. Vol. I. Addison Wesley, Reading, MA (1966)
12. Jarchow, H.: Locally Convex Spaces. Mathematical Textbooks. Teubner, Leipzig (1981)
13. Jost, J.: Postmodern Analysis, 3 edn. Universitext. Springer-Verlag, New York/Heidelberg/-Berlin (2005)
14. Kobayashi, S., Nomizu, K.: Foundations of Differential Geometry, Volume I. No. 15 in Interscience Tracts in Pure and Applied Mathematics. Interscience Publishers, New York (1963)
15. Kolář, I., Michor, P.W., Slovák, J.: Natural Operations in Differential Geometry. Springer-Verlag, New York/Heidelberg/Berlin (1993)
16. Michor, P.W.: Manifolds of Differentiable Mappings. No. 3 in Shiva Mathematics Series. Shiva Publishing Limited, Orpington, UK (1980)
17. Munkres, J.R.: Topology, 2 edn. Prentice-Hall, Englewood Cliffs, NJ (2000)
18. Pietsch, A.: Nuclear Locally Convex Spaces. No. 66 in Ergebnisse der Mathematik und ihrer Grenzgebiete. Springer-Verlag, New York/Heidelberg/Berlin (1969)

19. Robertson, A.P., Robertson, W.: Topological Vector Spaces, 2 edn. No. 53 in Cambridge Tracts in Mathematics. Cambridge University Press, New York/Port Chester/Melbourne/Sydney (1980)
20. Rudin, W.: Functional Analysis, 2 edn. International Series in Pure and Applied Mathematics. McGraw-Hill, New York (1991)
21. Sasaki, S.: On the differential geometry of tangent bundles of Riemannian manifolds. The Tôhoku Mathematical Journal. Second Series **10**, 338–354 (1958)
22. Schaefer, H.H., Wolff, M.P.: Topological Vector Spaces, 2 edn. No. 3 in Graduate Texts in Mathematics. Springer-Verlag, New York/Heidelberg/Berlin (1999)
23. Schwartz, L.: Radon Measures on Arbitrary Topological Spaces and Cylindrical Measures. Tata Institute Monographs on Mathematics & Physics. Oxford University Press, Oxford (1974)
24. Stein, E.M., Weiss, G.: Introduction to Fourier Analysis on Euclidean Space. No. 32 in Princeton Mathematical Series. Princeton University Press, Princeton, NJ (1971)
25. Sussmann, H.J.: An introduction to the coordinate-free maximum principle. In: B. Jakubczyk, W. Respondek (eds.) Geometry of Feedback and Optimal Control, pp. 463–557. Dekker Marcel Dekker, New York (1997)
26. Thomas, G.E.F.: Integration of functions with values in locally convex Suslin spaces. Transactions of the American Mathematical Society **212**, 61–81 (1975)
27. Weaver, N.: Lipschitz Algebras. World Scientific, Singapore/New Jersey/London/Hong Kong (1999)
28. Willard, S.: General Topology. Dover Publications, Inc., New York (2004). Reprint of 1970 Addison-Wesley edition

Chapter 4
The $\mathrm{CO}^{\mathrm{hol}}$-Topology for the Space of Holomorphic Vector Fields

Even if one has no per se interest in holomorphic vector fields, it is the case that an understanding of certain constructions for real analytic vector fields relies in an essential way on their holomorphic extensions. Also, as we shall see, we will arrive at a description of the real analytic topology that, while often easy to use in general arguments, is not well suited for verifying hypotheses in examples. In these cases, it is often most convenient to extend from real analytic to holomorphic, where things are easier to verify.

Thus in this section we overview the holomorphic case. We begin with vector bundles, as in the smooth case.

4.1 General Holomorphic Vector Bundles

We let $\pi\colon \mathsf{E} \to \mathsf{M}$ be an holomorphic vector bundle with $\Gamma^{\mathrm{hol}}(\mathsf{E})$ the set of holomorphic sections. We let $\mathbb{G}$ be an Hermitian fibre metric on E, and, for $K \subseteq \mathsf{M}$ compact, define a seminorm p_K^{hol} on $\Gamma^{\mathrm{hol}}(\mathsf{E})$ by

$$p_K^{\mathrm{hol}}(\xi) = \sup\{\|\xi(z)\|_{\mathbb{G}} \mid z \in K\}.$$

The $\mathbf{CO}^{\mathbf{hol}}$***-topology*** for $\Gamma^{\mathrm{hol}}(\mathsf{E})$ is the locally convex topology defined by the family of seminorms p_K^{hol}, $K \subseteq \mathsf{M}$ compact.

We shall have occasion to make use of bounded holomorphic sections. Thus we let $\pi\colon \mathsf{E} \to \mathsf{M}$ be an holomorphic vector bundle with Hermitian fibre metric $\mathbb{G}$. We denote by $\Gamma_{\mathrm{bdd}}^{\mathrm{hol}}(\mathsf{E})$ the sections of E that are bounded, and on $\Gamma_{\mathrm{bdd}}^{\mathrm{hol}}(\mathsf{E})$ we define a norm

$$p_{\infty}^{\mathrm{hol}}(\xi) = \sup\{\|\xi(z)\|_{\mathbb{G}} \mid z \in \mathsf{M}\}.$$

If we wish to draw attention to the domain of the section, we will write the norm as $p_{\mathsf{M},\infty}^{\mathrm{hol}}$. This will occur when we have sections defined on an open subset of the manifold.

S. Jafarpour, A.D. Lewis, *Time-Varying Vector Fields and Their Flows*,
SpringerBriefs in Mathematics, DOI 10.1007/978-3-319-10139-2_4

The following lemma makes an assertion of which we shall make use.

Lemma 4.1 (The Topology of $\Gamma^{hol}_{bdd}(\mathsf{E})$). *Let $\pi\colon \mathsf{E} \to \mathsf{M}$ be an holomorphic vector bundle. The subspace topology on $\Gamma^{hol}_{bdd}(\mathsf{E})$, induced from the CO^{hol}-topology, is weaker than the norm topology induced by the norm p^{hol}_{∞}. Moreover, $\Gamma^{hol}_{bdd}(\mathsf{E})$ is a Banach space. Also, if $\mathcal{U} \subseteq \mathsf{M}$ is a relatively compact open set with $\mathrm{cl}(\mathcal{U}) \subset \mathsf{M}$, then the restriction map from $\Gamma^{hol}(\mathsf{E})$ to $\Gamma^{hol}_{bdd}(\mathsf{E}|\mathcal{U})$ is continuous.*

Proof: It suffices to show that a sequence $(\xi_j)_{j\in\mathbb{Z}_{>0}}$ in $\Gamma^{hol}_{bdd}(\mathsf{E})$ converges to $\xi \in \Gamma^{hol}_{bdd}(\mathsf{E})$ uniformly on compact subsets of M if it converges in norm. This, however, is obvious. It remains to prove completeness of $\Gamma^{hol}_{bdd}(\mathsf{E})$ in the norm topology. By [4, Theorem 7.9], a Cauchy sequence $(\xi_j)_{j\in\mathbb{Z}_{>0}}$ in $\Gamma^{hol}_{bdd}(\mathsf{E})$ converges to a bounded continuous section ξ of E. That ξ is also holomorphic follows since uniform limits of holomorphic sections are holomorphic [3, p. 5]. For the final assertion, since the topology of $\Gamma^{hol}(\mathsf{E})$ is metrisable (see CO^{hol}-3 below), it suffices to show that the restriction of a convergent sequence in $\Gamma^{hol}(\mathsf{E})$ to $\mathcal{U}$ converges uniformly. This, however, follows since $\mathrm{cl}(\mathcal{U})$ is compact. □

One of the useful attributes of holomorphic geometry is that properties of higher derivatives can be deduced from the mapping itself. To make this precise, we first make the following observations.

1. Hermitian inner products on $\mathbb{C}$-vector spaces give inner products on the underlying $\mathbb{R}$-vector space.
2. By Lemma 2.4, there exist a real analytic affine connection ∇ on M and a real analytic vector bundle connection ∇^0 on E.
3. The estimates of Lemma 2.5 hold if the Riemannian metric $\mathbb{G}$ and the vector bundle metric $\mathbb{G}_0$ are only smooth. This is true because, in the proof, $\mathbb{G}$ and $\mathbb{G}_0$ are not differentiated; one only requires their values.

Therefore, the seminorms defined in Sect. 3.1 can be made sense of for holomorphic sections.

Proposition 4.2 (Cauchy Estimates for Vector Bundles). *Let $\pi\colon \mathsf{E} \to \mathsf{M}$ be an holomorphic vector bundle, let $K \subseteq \mathsf{M}$ be compact, and let $\mathcal{U}$ be a relatively compact neighbourhood of K. Then there exist $C, r \in \mathbb{R}_{>0}$ such that*

$$p^{\infty}_{K,m}(\xi) \le C r^{-m} p^{hol}_{\mathcal{U},\infty}(\xi)$$

for every $m \in \mathbb{Z}_{\ge 0}$ and $\xi \in \Gamma^{hol}_{bdd}(\mathsf{E}|\mathcal{U})$.

Moreover, if $(\mathcal{U}_j)_{j\in\mathbb{Z}_{>0}}$ is a sequence of relatively compact neighbourhoods of K such that (i) $\mathrm{cl}(\mathcal{U}_{j+1}) \subseteq \mathcal{U}_j$ and (ii) $K = \cap_{j\in\mathbb{Z}_{>0}} \mathcal{U}_j$, and if $C_j, r_j \in \mathbb{R}_{>0}$ are such that

$$p^{\infty}_{K,m}(\xi) \le C_j r_j^{-m} p^{hol}_{\mathcal{U}_j,\infty}(\xi), \qquad m \in \mathbb{Z}_{\ge 0},\ \xi \in \Gamma^{hol}_{bdd}(\mathsf{E}|\mathcal{U}_j),$$

then $\lim_{j\to\infty} r_j = 0$.

Proof: Let $z \in K$ and let $(\mathcal{W}_z, \psi_z)$ be an holomorphic vector bundle chart about z with $(\mathcal{U}_z, \phi_z)$ the associated chart for M, supposing that $\mathcal{U}_z \subseteq \mathcal{U}$. Let $k \in \mathbb{Z}_{>0}$ be such that $\psi_z(\mathcal{W}_z) = \phi_z(\mathcal{U}_z) \times \mathbb{C}^k$. Let $\boldsymbol{z} = \phi_z(z)$ and let $\boldsymbol{\xi}\colon \phi_z(\mathcal{U}_z) \to \mathbb{C}^k$ be the local representative of $\xi \in \Gamma^{\text{hol}}_{\text{bdd}}(\mathsf{E}|\mathcal{U})$. Note that when taking real derivatives of ξ with respect to coordinates, we can think of taking derivatives with respect to

$$\frac{\partial}{\partial z^j} = \frac{1}{2}\Big(\frac{\partial}{\partial x^j} - \mathrm{i}\frac{\partial}{\partial y^j}\Big), \quad \frac{\partial}{\partial \bar{z}^j} = \frac{1}{2}\Big(\frac{\partial}{\partial x^j} + \mathrm{i}\frac{\partial}{\partial y^j}\Big), \qquad j \in \{1,\dots,n\}.$$

Since ξ is holomorphic, the $\frac{\partial}{\partial \bar{z}^j}$ derivatives will vanish [7, p. 27]. Thus, for the purposes of the multi-index calculations, we consider multi-indices of length n (not $2n$). In any case, applying the usual Cauchy estimates [7, Lemma 2.3.9], there exists $r \in \mathbb{R}_{>0}$ such that

$$|\boldsymbol{D}^I \xi^a(z)| \le I!r^{-|I|} \sup\{|\xi^a(\zeta)| \mid \zeta \in \overline{\mathsf{D}}(\boldsymbol{r}, z)\}$$

for every $a \in \{1,\dots,k\}$, $I \in \mathbb{Z}^n_{\ge 0}$, and $\xi \in \Gamma^{\text{hol}}_{\text{bdd}}(\mathsf{E}|\mathcal{U})$. We may choose $r \in (0,1)$ such that $\overline{\mathsf{D}}(\boldsymbol{r}, z)$ is contained in $\phi_z(\mathcal{U}_z)$, where $\boldsymbol{r} = (r,\dots,r)$. Denote $\mathcal{V}_z = \phi_z^{-1}(\mathsf{D}(\boldsymbol{r}, z))$. There exists a neighbourhood $\mathcal{V}'_z$ of z such that $\mathrm{cl}(\mathcal{V}'_z) \subseteq \mathcal{V}_z$ and such that

$$|\boldsymbol{D}^I \xi^a(z')| \le 2I!r^{-|I|} \sup\{|\xi^a(\zeta)| \mid \zeta \in \overline{\mathsf{D}}(\boldsymbol{r}, z)\}$$

for every $z' \in \phi_z(\mathcal{V}'_z)$, $\xi \in \Gamma^{\text{hol}}_{\text{bdd}}(\mathsf{E}|\mathcal{U})$, $a \in \{1,\dots,k\}$, and $I \in \mathbb{Z}^n_{\ge 0}$. If $|I| \le m$, then, since we are assuming that $r < 1$, we have

$$\frac{1}{I!}|\boldsymbol{D}^I \xi^a(z')| \le 2r^{-m} \sup\{|\xi^a(\zeta)| \mid \zeta \in \overline{\mathsf{D}}(\boldsymbol{r}, z)\}$$

for every $a \in \{1,\dots,k\}$, $z' \in \phi_z(\mathcal{V}'_z)$, and $\xi \in \Gamma^{\text{hol}}_{\text{bdd}}(\mathsf{E}|\mathcal{U})$. By Lemma 2.5 (and keeping in mind our observations made before the statement of the proposition), it follows that there exist $C_z, r_z \in \mathbb{R}_{>0}$ such that

$$\|j_m\xi(z)\|_{\overline{\mathbb{G}}_m} \le C_z r_z^{-m} p^{\text{hol}}_{\mathcal{V}_z,\infty}(\xi)$$

for all $z \in \mathcal{V}'_z$, $m \in \mathbb{Z}_{\ge 0}$, and $\xi \in \Gamma^{\text{hol}}_{\text{bdd}}(\mathsf{E}|\mathcal{U})$. Let $z_1,\dots,z_k \in K$ be such that $K \subseteq \cup_{j=1}^k \mathcal{V}'_{z_j}$, and let $C = \max\{C_{z_1},\dots,C_{z_k}\}$ and $r = \min\{r_{z_1},\dots,r_{z_k}\}$. If $z \in K$, then $z \in \mathcal{V}'_{z_j}$ for some $j \in \{1,\dots,k\}$ and so we have

$$\|j_m\xi(z)\|_{\overline{\mathbb{G}}_m} \le C_{z_j} r_{z_j}^{-m} p^{\text{hol}}_{\mathcal{V}_{z_j},\infty}(\xi) \le Cr^{-m} p^{\text{hol}}_{\mathcal{U},\infty}(\xi),$$

and taking supremums over $z \in K$ on the left gives the result.

The final assertion of the proposition immediately follows by observing in the preceding construction how "r" was defined, namely that it had to be chosen so that polydisks of radius r in the coordinate charts remained in $\mathcal{U}$. □

4.2 Properties of the CO^{hol}-Topology

The CO^{hol}-topology for $\Gamma^{hol}(E)$ has the following attributes.

CO^{hol}-1. It is Hausdorff: [8, Theorem 8.2].

CO^{hol}-2. It is complete: [8, Theorem 8.2].

CO^{hol}-3. It is metrisable: [8, Theorem 8.2].

CO^{hol}-4. It is separable: This follows since $\Gamma^{hol}(E)$ is a closed subspace of $\Gamma^{\infty}(E)$ by [8, Theorem 8.2] and since subspaces of separable metric spaces are separable [12, Theorems 16.2, 16.9, and 16.11].

CO^{hol}-5. It is nuclear: [8, Theorem 8.2]. Note that, when M is compact, p_{M}^{hol} is a norm for the C^{hol}-topology. A consequence of this is that $\Gamma^{hol}(E)$ must be finite-dimensional in these cases since the only nuclear normed vector spaces are those that are finite-dimensional [9, Theorem 4.4.14].

CO^{hol}-6. It is Suslin: This follows since $\Gamma^{hol}(E)$ is a Polish space, as we have seen above.

Being metrisable, it suffices to describe the CO^{hol}-topology by describing its convergent sequences; these are more or less obviously the sequences that converge uniformly on every compact set.

As with spaces of smooth sections, we are interested in the fact that nuclearity of $\Gamma^{hol}(E)$ implies that compact sets are exactly those sets that are closed and von Neumann bounded. The following result is obvious in the same way that Lemma 3.1 is obvious once one understands Theorem 1.37(b) from [10].

Lemma 4.3 (Bounded Subsets in the CO^{hol}-Topology). *A subset $\mathcal{B} \subseteq \Gamma^{hol}(E)$ is bounded in the von Neumann bornology if and only if the following property holds: for any compact set $K \subseteq M$, there exists $C \in \mathbb{R}_{>0}$ such that $p_{K}^{hol}(\xi) \leq C$ for every $\xi \in \mathcal{B}$.*

4.3 The Weak-$\mathscr{L}$ Topology for Holomorphic Vector Fields

As in the smooth case, one simply specialises the constructions for general vector bundles to get the CO^{hol}-topology for the space $\Gamma^{hol}(TM)$ of holomorphic vector fields and the space $C^{hol}(M)$ of holomorphic functions, noting that an holomorphic function is obviously identified with a section of the trivial holomorphic vector bundle $M \times \mathbb{C}$.

As with smooth vector fields, for holomorphic vector fields we can seek a weak-$\mathscr{L}$ characterisation of the CO^{hol}-topology. To begin, we need to understand the Lie derivative in the holomorphic case. Thinking of $C^{hol}(M) \subseteq C^{\infty}(M) \otimes \mathbb{C}$ and using the Wirtinger formulae,

$$\frac{\partial}{\partial z^j} = \frac{1}{2}\Big(\frac{\partial}{\partial x^j} - \mathrm{i}\frac{\partial}{\partial y^j}\Big), \quad \frac{\partial}{\partial \bar{z}^j} = \frac{1}{2}\Big(\frac{\partial}{\partial x^j} + \mathrm{i}\frac{\partial}{\partial y^j}\Big), \qquad j \in \{1, \dots, n\},$$

in an holomorphic chart, one sees that the usual differential of a $\mathbb{C}$-valued function can be decomposed as $\mathrm{d}_{\mathbb{C}} f = \partial f + \bar{\partial} f$, the first term on the right corresponding to "$\frac{\partial}{\partial z}$" and the second to "$\frac{\partial}{\partial \bar{z}}$". For holomorphic functions, the Cauchy–Riemann equations [7, p. 27] imply that $\mathrm{d}_{\mathbb{C}} f = \partial f$. Thus we define the Lie derivative of an holomorphic function f with respect to an holomorphic vector field X by $\mathscr{L}_X f = \langle \partial f; X \rangle$. Fortunately, in coordinates this assumes the expected form:

$$\mathscr{L}_X f = \sum_{j=1}^{n} X^j \frac{\partial f}{\partial z^j}.$$

It is *not* the case that on a general holomorphic manifold there is a correspondence between derivations of the $\mathbb{C}$-algebra $\mathrm{C}^{\mathrm{hol}}(\mathsf{M})$ and holomorphic vector fields by Lie differentiation.[1] However, for a certain class of holomorphic manifolds, those known as "Stein manifolds", the exact correspondence between derivations of the $\mathbb{C}$-algebra $\mathrm{C}^{\mathrm{hol}}(\mathsf{M})$ and holomorphic vector fields under Lie differentiation *does* hold [2]. This is good news for us, since Stein manifolds are intimately connected with real analytic manifolds, as we shall see in the next section.

With the preceding discussion in mind, we can move ahead with Definition 3.2 with $\mathsf{U} = \Gamma^{\mathrm{hol}}(\mathsf{TM})$, $\mathsf{V} = \mathrm{C}^{\mathrm{hol}}(\mathsf{M})$ (with the $\mathrm{CO}^{\mathrm{hol}}$-topology), and $\mathscr{A} = \{\mathscr{L}_f \mid f \in \mathrm{C}^{\mathrm{hol}}(\mathsf{M})\}$, where

$$\begin{aligned} \mathscr{L}_f \colon \Gamma^{\mathrm{hol}}(\mathsf{TM}) &\to \mathrm{C}^{\mathrm{hol}}(\mathsf{M}) \\ X &\mapsto \mathscr{L}_X f. \end{aligned}$$

We make the following definition.

Definition 4.4 (Weak-$\mathscr{L}$ Topology for Space of Holomorphic Vector Fields). For an holomorphic manifold M, the ***weak-$\mathscr{L}$ topology*** for $\Gamma^{\mathrm{hol}}(\mathsf{TM})$ is the weakest topology for which $\mathscr{L}_f$ is continuous for every $f \in \mathrm{C}^{\mathrm{hol}}(\mathsf{M})$, if $\mathrm{C}^{\mathrm{hol}}(\mathsf{M})$ has the $\mathrm{CO}^{\mathrm{hol}}$-topology. ∘

We then have the following result.

Theorem 4.5 (Weak-$\mathscr{L}$ Characterisation of $\mathrm{CO}^{\mathrm{hol}}$-Topology for Holomorphic Vector Fields on Stein Manifolds). *For a Stein manifold* M, *the following topologies for* $\Gamma^{\mathrm{hol}}(\mathsf{TM})$ *agree:*

(i) the $\mathrm{CO}^{\mathrm{hol}}$*-topology;*

(ii) the weak-$\mathscr{L}$ topology.

Proof: (i)⊆(ii) As we argued in the proof of the corresponding assertion of (3.5), it suffices to show that

$$p_K^{\mathrm{hol}}(X) \le C_1 p_{K_1}^{\mathrm{hol}}(\mathscr{L}_X f^1) + \cdots + C_r p_{K_r}^{\mathrm{hol}}(\mathscr{L}_X f^r)$$

for some $C_1, \ldots, C_r \in \mathbb{R}_{>0}$, some $K_1, \ldots, K_r \subseteq \mathsf{M}$ compact, and some $f^1, \ldots, f^r \in \mathrm{C}^{\mathrm{hol}}(\mathsf{M})$.

[1] For example, on a compact holomorphic manifold, the only holomorphic functions are locally constant [1, Corollary IV.1.3], and so the only derivation is the zero derivation. However, the $\mathbb{C}$-vector space of holomorphic vector fields, while not large, may have positive dimension. For example, the space of holomorphic vector fields on the Riemann sphere has $\mathbb{C}$-dimension three [6, Problem 17.9].

Let $K \subseteq \mathsf{M}$ be compact. For simplicity, we assume that M is connected and so has a well-defined dimension n. If not, then the arguments are easily modified by change of notation to account for this. Since M is a Stein manifold, for every $z \in K$ there exists a coordinate chart $(\mathcal{U}_z, \phi_z)$ with coordinate functions $z^1, \dots, z^n : \mathcal{U}_z \to \mathbb{C}$ that are restrictions to $\mathcal{U}_z$ of globally defined holomorphic functions on M. Depending on your source, this is either a theorem or part of the definition of a Stein manifold [1, 5]. Thus, for $j \in \{1, \dots, n\}$, let $f_z^j \in \mathrm{C}^{\mathrm{hol}}(\mathsf{M})$ be the holomorphic function which, when restricted to $\mathcal{U}_z$, gives the coordinate function z^j. Clearly, $\mathscr{L}_X f_z^j = X^j$ on $\mathcal{U}_z$. Also, there exists $C_z \in \mathbb{R}_{>0}$ such that

$$\|X(\zeta)\|_{\mathbb{G}} \leq C_z(|X^1(\zeta)| + \cdots + |X^n(\zeta)|), \qquad \zeta \in \mathrm{cl}(\mathcal{V}_z),$$

for some relatively compact neighbourhood $\mathcal{V}_z \subseteq \mathcal{U}_z$ of z (this follows from the fact that all norms are equivalent to the ℓ^1 norm for $\mathbb{C}^n$). Thus

$$\|X(\zeta)\|_{\mathbb{G}} \leq C_z(|\mathscr{L}_X f_z^1(\zeta)| + \cdots + |\mathscr{L}_X f_z^n(\zeta)|), \qquad \zeta \in \mathrm{cl}(\mathcal{V}_z).$$

Let $z_1, \dots, z_k \in K$ be such that $K \subseteq \cup_{j=1}^k \mathcal{V}_{z_j}$. Let $f^1, \dots, f^{kn}$ be the list of globally defined holomorphic functions

$$f_{z_1}^1, \dots, f_{z_1}^n, \dots, f_{z_k}^1, \dots, f_{z_k}^n$$

and let $C_1, \dots, C_{kn}$ be the list of coefficients

$$\underbrace{C_{z_1}, \dots, C_{z_1}}_{n \text{ times}}, \dots, \underbrace{C_{z_k}, \dots, C_{z_k}}_{n \text{ times}}.$$

If $z \in K$, then $z \in \mathcal{V}_{z_j}$ for some $j \in \{1, \dots, k\}$ and so

$$\|X(z)\|_{\mathbb{G}} \leq C_1|\mathscr{L}_X f^1(z)| + \cdots + C_{kn}|\mathscr{L}_X f^{kn}(z)|,$$

which gives

$$p_K^{\mathrm{hol}}(X) \leq C_1 p_K^{\mathrm{hol}}(\mathscr{L}_X f^1) + \cdots + C_{kn} p_K^{\mathrm{hol}}(\mathscr{L}_X f^{kn}),$$

as needed.

(ii)$\subseteq$(i) We claim that $\mathscr{L}_f$ is continuous for every $f \in \mathrm{C}^{\mathrm{hol}}(\mathsf{M})$ if $\Gamma^{\mathrm{hol}}(\mathsf{TM})$ has the $\mathrm{CO^{hol}}$-topology. Let $K \subseteq \mathsf{M}$ be compact and let $\mathcal{U}$ be a relatively compact neighbourhood of K in M. Note that, for $f \in \mathrm{C}^{\mathrm{hol}}(\mathsf{M})$,

$$p_K^{\mathrm{hol}}(\mathscr{L}_X f) \leq C p_{K,1}^{\infty}(f) p_K^{\mathrm{hol}}(X) \leq C' p_K^{\mathrm{hol}}(X),$$

using Proposition 4.2, giving continuity of the identity map if we provide the domain with the $\mathrm{CO^{hol}}$-topology and the codomain with the weak-$\mathscr{L}$ topology, cf. [11, Sect. III.1.1]. Thus open sets in the weak-$\mathscr{L}$ topology are contained in the $\mathrm{CO^{hol}}$-topology. □

As in the smooth case, we shall use the theorem according to the following result.

Corollary 4.6 (Weak-$\mathscr{L}$ Characterisations of Boundedness, Continuity, Measurability, and Integrability for the $\mathrm{CO}^{\mathrm{hol}}$-Topology). *Let* M *be a Stein manifold, let* $(\mathcal{X}, \mathscr{O})$ *be a topological space, let* $(\mathcal{T}, \mathscr{M})$ *be a measurable space, and let* $\mu\colon \mathscr{M} \to \overline{\mathbb{R}}_{\geq 0}$ *be a finite measure. The following statements hold:*

(i) a subset $\mathcal{B} \subseteq \Gamma^{\mathrm{hol}}(\mathsf{TM})$ *is bounded in the von Neumann bornology if and only if it is weak-*$\mathscr{L}$ *bounded in the von Neumann bornology;*
(ii) a map $\Phi\colon \mathcal{X} \to \Gamma^{\mathrm{hol}}(\mathsf{TM})$ *is continuous if and only if it is weak-*$\mathscr{L}$ *continuous;*
(iii) a map $\Psi\colon \mathcal{T} \to \Gamma^{\mathrm{hol}}(\mathsf{TM})$ *is measurable if and only if it is weak-*$\mathscr{L}$ *measurable;*
(iv) a map $\Psi\colon \mathcal{T} \to \Gamma^{\mathrm{hol}}(\mathsf{TM})$ *is Bochner integrable if and only if it is weak-*$\mathscr{L}$ *Bochner integrable.*

Proof: As in the proof of Corollary 3.6, we need to show that $\{\mathscr{L}_f \mid f \in \mathrm{C}^{\mathrm{hol}}(\mathsf{M})\}$ has a countable point separating subset. The argument here follows that in the smooth case, except that here we have to use the properties of Stein manifolds, cf. the proof of the first part of Theorem 4.5 above, to assert the existence, for each $z \in \mathsf{M}$, of a neighbourhood on which there are globally defined holomorphic functions whose differentials span the cotangent space at each point. Since $\Gamma^{\mathrm{hol}}(\mathsf{TM})$ is complete, separable, and Suslin, and since $\mathrm{C}^{\mathrm{hol}}(\mathsf{M})$ is Suslin by properties $\mathrm{CO}^{\mathrm{hol}}$-2, $\mathrm{CO}^{\mathrm{hol}}$-4 and $\mathrm{CO}^{\mathrm{hol}}$-6 above, the corollary follows from Lemma 3.3. □

References

1. Fritzsche, K., Grauert, H.: From Holomorphic Functions to Complex Manifolds. No. 213 in Graduate Texts in Mathematics. Springer-Verlag, New York/Heidelberg/Berlin (2002)
2. Grabowski, J.: Derivations of Lie algebras of analytic vector fields. Compositio Mathematica **43**(2), 239–252 (1981)
3. Gunning, R.C.: Introduction to Holomorphic Functions of Several Variables. Volume I: Function Theory. Wadsworth & Brooks/Cole Mathematics Series. Wadsworth & Brooks/Cole, Belmont, CA (1990)
4. Hewitt, E., Stromberg, K.: Real and Abstract Analysis. No. 25 in Graduate Texts in Mathematics. Springer-Verlag, New York/Heidelberg/Berlin (1975)
5. Hörmander, L.: An Introduction to Complex Analysis in Several Variables, 2 edn. North-Holland, Amsterdam/New York (1973)
6. Ilyashenko, Y., Yakovenko, S.: Lectures on Analytic Differential Equations. No. 86 in Graduate Studies in Mathematics. American Mathematical Society, Providence, RI (2008)
7. Krantz, S.G.: Function Theory of Several Complex Variables, 2 edn. AMS Chelsea Publishing, Providence, RI (1992)
8. Kriegl, A., Michor, P.W.: The Convenient Setting of Global Analysis. No. 57 in American Mathematical Society Mathematical Surveys and Monographs. American Mathematical Society, Providence, RI (1997)
9. Pietsch, A.: Nuclear Locally Convex Spaces. No. 66 in Ergebnisse der Mathematik und ihrer Grenzgebiete. Springer-Verlag, New York/Heidelberg/Berlin (1969)
10. Rudin, W.: Functional Analysis, 2 edn. International Series in Pure and Applied Mathematics. McGraw-Hill, New York (1991)
11. Schaefer, H.H., Wolff, M.P.: Topological Vector Spaces, 2 edn. No. 3 in Graduate Texts in Mathematics. Springer-Verlag, New York/Heidelberg/Berlin (1999)
12. Willard, S.: General Topology. Dover Publications, Inc., New York (2004). Reprint of 1970 Addison-Wesley edition

Chapter 5
The C^ω-Topology for the Space of Real Analytic Vector Fields

In this chapter we examine a topology on the set of real analytic vector fields. As we shall see, this requires some considerable effort. Agrachev and Gamkrelidze [1] consider the real analytic case by considering bounded holomorphic extensions to neighbourhoods of $\mathbb{R}^n$ of fixed width in $\mathbb{C}^n$. Our approach is more general, more geometric, and global, using a natural real analytic topology described, for example, in the work of Martineau [17]. This allows us to dramatically broaden the class of real analytic vector fields that we can handle to include "all" analytic vector fields.

The first observation we make is that $\Gamma^\omega(\mathsf{E})$ is not a closed subspace of $\Gamma^\infty(\mathsf{E})$ in the CO^∞-topology. To see this, consider the following. Take a smooth but not real analytic function on $\mathbb{S}^1$. The Fourier series of this function gives rise, by taking partial sums, to a sequence of real analytic functions. Standard harmonic analysis [23, Theorem VII.2.11(b)] shows that this sequence and all of its derivatives converge uniformly, and so in the CO^∞-topology, to the original function. Thus we have a Cauchy sequence in $\mathrm{C}^\omega(\mathbb{S}^1)$ that does not converge, with respect to the CO^∞-topology, in $\mathrm{C}^\omega(\mathbb{S}^1)$.

The second observation we make is that a plain restriction of the topology for holomorphic objects is not sufficient. The reason for this is that, upon complexification (a process we describe in detail below) there will not be a uniform neighbourhood to which all real analytic objects can be extended. Let us look at this for an example, where "object" is "function". For $r \in \mathbb{R}_{>0}$ we consider the real analytic function $f_r \colon \mathbb{R} \to \mathbb{R}$ defined by $f_r(x) = \frac{r^2}{r^2+x^2}$. We claim that there is no neighbourhood $\overline{\mathcal{U}}$ of $\mathbb{R}$ in $\mathbb{C}$ to which all of the functions f_r, $r \in \mathbb{R}_{>0}$, can be extended. Indeed, take some such neighbourhood $\overline{\mathcal{U}}$ and let $r \in \mathbb{R}_{>0}$ be sufficiently small that $\overline{\mathsf{D}}(r,0) \subseteq \overline{\mathcal{U}}$. To see that f_r cannot be extended to an holomorphic function on $\overline{\mathcal{U}}$, let $\overline{f}_r$ be such an holomorphic extension. Then $\overline{f}_r(z)$ must be equal to $\frac{r^2}{r^2+z^2}$ for $z \in \mathsf{D}(r,0)$ by uniqueness of holomorphic extensions [6, Lemma 5.40]. But this immediately prohibits $\overline{f}_r$ from being holomorphic on any neighbourhood of $\overline{\mathsf{D}}(r,0)$, giving our claim.

S. Jafarpour, A.D. Lewis, *Time-Varying Vector Fields and Their Flows*, SpringerBriefs in Mathematics, DOI 10.1007/978-3-319-10139-2_5

Therefore, to topologise the space of real analytic vector fields, we will need to do more than either (1) restrict the CO^∞-topology or (2) use the CO^{hol}-topology in an "obvious" way. Note that it is the "obvious" use of the CO^{hol}-topology for holomorphic objects that is employed by Agrachev and Gamkrelidze [1] in their study of time-varying real analytic vector fields. Moreover, Agrachev and Gamkrelidze also restrict to *bounded* holomorphic extensions. What we propose is an improvement on this in that it works far more generally and is also more natural to a geometric treatment of the real analytic setting. We comment at this point that we shall see in Theorem 6.25 below that the consideration of bounded holomorphic extensions to fixed neighbourhoods in the complexification is sometimes sufficient locally. But conclusions such as this become hard theorems with precise hypotheses in our approach, not starting points for the theory.

As in the smooth and holomorphic cases, we begin by considering a general vector bundle.

5.1 A Natural Direct Limit Topology

We let $\pi\colon \mathsf{E} \to \mathsf{M}$ be a real analytic vector bundle, and consider a fairly natural topology for the space $\Gamma^\omega(\mathsf{E})$ of real analytic sections. First we shall extend E to an holomorphic vector bundle that will serve an important device for all of our constructions.

5.1.1 *Complexifications*

Let us take some time to explain how holomorphic extensions can be constructed. The following two paragraphs distill out important parts of about 40 years of intensive development of complex analysis, culminating in the paper of Grauert [11].

For simplicity, let us assume that M is connected and so has pure dimension, and so the fibres of E also have a fixed dimension. As in Sect. 2.3, we suppose that we have a real analytic affine connection ∇ on M, a real analytic vector bundle connection ∇^0 on E, a real analytic Riemannian metric $\mathbb{G}$ on M, and a real analytic fibre metric $\mathbb{G}_0$ on E. We also assume the data required to make the diagram (2.7), giving $\pi\colon \mathsf{E} \to \mathsf{M}$ as the image of a real analytic vector bundle monomorphism in the trivial vector bundle $\mathbb{R}^N \times \mathbb{R}^N$ for some suitable $N \in \mathbb{Z}_{>0}$.

Now we complexify. Recall that, if V is a $\mathbb{C}$-vector space, then multiplication by $\sqrt{-1}$ induces a $\mathbb{R}$-linear map $J \in \mathrm{End}_{\mathbb{R}}(\mathsf{V})$. A $\mathbb{R}$-subspace U of V is ***totally real*** if $\mathsf{U} \cap J(\mathsf{U}) = \{0\}$. A smooth submanifold of an holomorphic manifold, thinking of the latter as a smooth manifold, is ***totally real*** if its tangent spaces are totally real subspaces. By [25, Proposition 1], for a real analytic manifold M there exists a complexification $\overline{\mathsf{M}}$ of M, i.e., an holomorphic manifold having M as a totally real submanifold and where $\overline{\mathsf{M}}$ has the same $\mathbb{C}$-dimension as the $\mathbb{R}$-dimension of M.

As shown by Grauert [11, Sect. 3.4], for any neighbourhood $\overline{\mathcal{U}}$ of M in $\overline{\mathsf{M}}$, there exists a Stein neighbourhood $\overline{\mathcal{S}}$ of M contained in $\overline{\mathcal{U}}$. By arguments involving extending convergent real power series to convergent complex power series (the conditions on coefficients for convergence are the same for both real and complex power series), one can show that there is an holomorphic extension of ι_{M} to $\iota_{\overline{\mathsf{M}}}\colon \overline{\mathsf{M}} \to \mathbb{C}^N$, possibly after shrinking $\overline{\mathsf{M}}$ [6, Lemma 5.40]. By applying similar reasoning to the transition maps for the real analytic vector bundle E, one obtains an holomorphic vector bundle $\overline{\pi}\colon \overline{\mathsf{E}} \to \overline{\mathsf{M}}$ for which the diagram

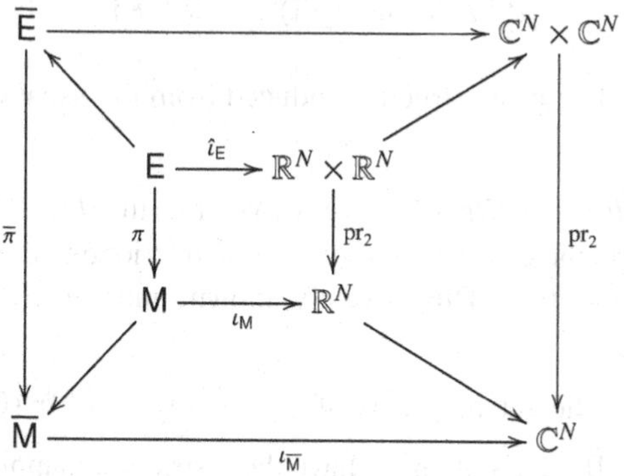

commutes, where all diagonal arrows are complexification and where the inner diagram is as defined in the proof of Lemma 2.4. One can then define an Hermitian fibre metric $\overline{\mathbb{G}}_0$ on $\overline{\mathsf{E}}$ induced from the standard Hermitian metric on the fibres of the vector bundle $\mathbb{C}^N \times \mathbb{C}^N$ and an Hermitian metric $\overline{\mathbb{G}}$ on $\overline{\mathsf{M}}$ induced from the standard Hermitian metric on $\mathbb{C}^N$.

In the remainder of this section, we assume that the preceding constructions have been done and fixed once and for all.

5.1.2 *Germs of Holomorphic Sections over Subsets of a Real Analytic Manifold*

In two different places, we will need to consider germs of holomorphic sections. In this section we organise the methodology for doing this to unify the notation.

Let $A \subseteq \mathsf{M}$ and let $\mathscr{N}_A$ be the set of neighbourhoods of A in the complexification $\overline{\mathsf{M}}$. For $\overline{\mathcal{U}}, \overline{\mathcal{V}} \in \mathscr{N}_A$, and for $\overline{\xi} \in \Gamma^{\mathrm{hol}}(\overline{\mathsf{E}}|\overline{\mathcal{U}})$ and $\overline{\eta} \in \Gamma^{\mathrm{hol}}(\overline{\mathsf{E}}|\overline{\mathcal{V}})$, we say that $\overline{\xi}$ is ***equivalent*** to $\overline{\eta}$ if there exist $\overline{\mathcal{W}} \in \mathscr{N}_A$ and $\overline{\zeta} \in \Gamma^{\mathrm{hol}}(\overline{\mathsf{E}}|\overline{\mathcal{W}})$ such that $\overline{\mathcal{W}} \subseteq \overline{\mathcal{U}} \cap \overline{\mathcal{V}}$ and such that

$$\overline{\xi}|\overline{\mathcal{W}} = \overline{\eta}|\overline{\mathcal{W}} = \overline{\zeta}.$$

By $\mathscr{G}^{\mathrm{hol}}_{A,\overline{\mathsf{E}}}$ we denote the set of equivalence classes, which we call the set of germs of sections of $\overline{\mathsf{E}}$ over A. By $[\overline{\xi}]_A$ we denote the equivalence class of $\overline{\xi} \in \Gamma^{\mathrm{hol}}(\overline{\mathsf{E}}|\overline{\mathcal{U}})$ for some $\overline{\mathcal{U}} \in \mathscr{N}_A$.

Now, for $x \in \mathsf{M}$, E_x is a totally real subspace of $\overline{\mathsf{E}}_x$ with half the real dimension, and so it follows that

$$\overline{\mathsf{E}}_x = \mathsf{E}_x \oplus J(\mathsf{E}_x),$$

where J is the complex structure on the fibres of $\overline{\mathsf{E}}$. For $\overline{\mathcal{U}} \in \mathscr{N}_A$, denote by $\Gamma^{\mathrm{hol},\mathbb{R}}(\overline{\mathsf{E}}|\overline{\mathcal{U}})$ those holomorphic sections $\bar{\xi}$ of $\overline{\mathsf{E}}|\overline{\mathcal{U}}$ such that $\bar{\xi}(x) \in \mathsf{E}_x$ for $x \in \overline{\mathcal{U}} \cap \mathsf{M}$. We think of this as being a locally convex topological $\mathbb{R}$-vector space with the seminorms $p^{\mathrm{hol}}_{\overline{K}}$, $\overline{K} \subseteq \overline{\mathcal{U}}$ compact, defined by

$$p^{\mathrm{hol}}_{\overline{K}}(\bar{\xi}) = \sup\{\|\bar{\xi}(\bar{x})\|_{\overline{\mathbb{G}}_0} \mid \bar{x} \in \overline{K}\},$$

i.e., we use the locally convex structure induced from the usual $\mathrm{CO}^{\mathrm{hol}}$-topology on $\Gamma^{\mathrm{hol}}(\overline{\mathsf{E}}|\overline{\mathcal{U}})$.

Remark 5.1 (Closedness of "Real" Sections). We note that $\Gamma^{\mathrm{hol},\mathbb{R}}(\overline{\mathsf{E}}|\overline{\mathcal{U}})$ is a closed $\mathbb{R}$-subspace of $\Gamma^{\mathrm{hol}}(\overline{\mathsf{E}})$ in the $\mathrm{CO}^{\mathrm{hol}}$-topology, i.e., the restriction of requiring "realness" on M is a closed condition. This is easily shown, and we often assume it often without mention. ∘

Denote by $\mathscr{G}^{\mathrm{hol},\mathbb{R}}_{A,\overline{\mathsf{E}}}$ the set of germs of sections from $\Gamma^{\mathrm{hol},\mathbb{R}}(\overline{\mathsf{E}}|\overline{\mathcal{U}})$, $\overline{\mathcal{U}} \in \mathscr{N}_A$. If $\overline{\mathcal{U}}_1, \overline{\mathcal{U}}_2 \in \mathscr{N}_A$ satisfy $\overline{\mathcal{U}}_1 \subseteq \overline{\mathcal{U}}_2$, then we have the restriction mapping

$$\begin{aligned} r_{\overline{\mathcal{U}}_2,\overline{\mathcal{U}}_1} \colon \Gamma^{\mathrm{hol},\mathbb{R}}(\overline{\mathsf{E}}|\overline{\mathcal{U}}_2) &\to \Gamma^{\mathrm{hol},\mathbb{R}}(\overline{\mathsf{E}}|\overline{\mathcal{U}}_1) \\ \bar{\xi} &\mapsto \bar{\xi}|\overline{\mathcal{U}}_1. \end{aligned}$$

This restriction is continuous since, for any compact set $\overline{K} \subseteq \overline{\mathcal{U}}_1 \subseteq \overline{\mathcal{U}}_2$ and any $\bar{\xi} \in \Gamma^{\mathrm{hol},\mathbb{R}}(\overline{\mathsf{E}}|\overline{\mathcal{U}}_2)$, we have $p^{\mathrm{hol}}_{\overline{K}}(r_{\overline{\mathcal{U}}_2,\overline{\mathcal{U}}_1}(\bar{\xi})) \leq p^{\mathrm{hol}}_{\overline{K}}(\bar{\xi})$ (in fact we have equality, but the inequality emphasises what is required for our assertion to be true [22, Sect. III.1.1]). We also have maps

$$\begin{aligned} r_{\overline{\mathcal{U}},A} \colon \Gamma^{\mathrm{hol},\mathbb{R}}(\overline{\mathsf{E}}|\overline{\mathcal{U}}) &\to \mathscr{G}^{\mathrm{hol},\mathbb{R}}_{A,\overline{\mathsf{E}}} \\ \bar{\xi} &\mapsto [\bar{\xi}]_A. \end{aligned}$$

Note that $\mathscr{N}_A$ is a directed set by inclusion; that is, $\overline{\mathcal{U}}_2 \preceq \overline{\mathcal{U}}_1$ if $\overline{\mathcal{U}}_1 \subseteq \overline{\mathcal{U}}_2$. Thus we have the directed system $(\Gamma^{\mathrm{hol},\mathbb{R}}(\mathsf{T}\overline{\mathcal{U}}))_{\overline{\mathcal{U}}\in\mathscr{N}_A}$, along with the mappings $r_{\overline{\mathcal{U}}_2,\overline{\mathcal{U}}_1}$, in the category of locally convex topological $\mathbb{R}$-vector spaces. The usual notion of direct limit in the category of $\mathbb{R}$-vector spaces gives $\mathscr{G}^{\mathrm{hol},\mathbb{R}}_{A,\overline{\mathsf{E}}}$, along with the linear mappings $r_{\overline{\mathcal{U}},A}$, $\overline{\mathcal{U}} \in \mathscr{N}_A$, as the direct limit of this directed system, cf. [16, Theorem III.10.1]. This vector space then has the finest locally convex topology making the maps $r_{\overline{\mathcal{U}},A}$, $\overline{\mathcal{U}} \in \mathscr{N}_A$, continuous, i.e., the direct limit in the category of locally convex topological vector spaces. We refer to this as the ***direct limit topology*** for $\mathscr{G}^{\mathrm{hol},\mathbb{R}}_{A,\overline{\mathsf{E}}}$.

5.1.3 The Direct Limit Topology

We shall describe four topologies (or more, depending on which descriptions you regard as being distinct) for the space of real analytic sections of a real analytic vector bundle. The first is quite direct, involving an application of the construction above to the case of $A = \mathsf{M}$. In this case, the following lemma is key to our constructions.

Lemma 5.2 (Real Analytic Sections as Holomorphic Germs). *There is a natural* $\mathbb{R}$*-vector space isomorphism between* $\Gamma^{\omega}(\mathsf{E})$ *and* $\mathscr{G}^{\mathrm{hol},\mathbb{R}}_{\mathsf{M},\overline{\mathsf{E}}}$.

Proof: Let $\xi \in \Gamma^{\omega}(\mathsf{E})$. As in [6, Lemma 5.40], there is an extension of ξ to a section $\overline{\xi} \in \Gamma^{\mathrm{hol},\mathbb{R}}(\overline{\mathsf{E}}|\overline{\mathcal{U}})$ for some $\overline{\mathcal{U}} \in \mathscr{N}_{\mathsf{M}}$. We claim that the map $i_{\mathsf{M}} \colon \Gamma^{\omega}(\mathsf{E}) \to \mathscr{G}^{\mathrm{hol},\mathbb{R}}_{\mathsf{M},\overline{\mathsf{E}}}$ defined by $i_{\mathsf{M}}(\xi) = [\overline{\xi}]_{\mathsf{M}}$ is the desired isomorphism. That i_{M} is independent of the choice of extension $\overline{\xi}$ is a consequence of the fact that the extension to $\overline{\xi}$ is unique inasmuch as any two such extensions agree on some neighbourhood contained in their intersection; this is the uniqueness assertion of [6, Lemma 5.40]. This fact also ensures that i_{M} is injective. For surjectivity, let $[\overline{\xi}]_{\mathsf{M}} \in \mathscr{G}^{\mathrm{hol},\mathbb{R}}_{\mathsf{M},\overline{\mathsf{E}}}$ and let us define $\xi \colon \mathsf{M} \to \mathsf{E}$ by $\xi(x) = \overline{\xi}(x)$ for $x \in \mathsf{M}$. Note that the restriction of $\overline{\xi}$ to M is real analytic because the values of $\overline{\xi}|\mathsf{M}$ at points in a neighbourhood of $x \in \mathsf{M}$ are given by the restriction of the (necessarily convergent) $\mathbb{C}$-Taylor series of $\overline{\xi}$ to M. Obviously, $i_{\mathsf{M}}(\xi) = [\overline{\xi}]_{\mathsf{M}}$. □

Now we use the direct limit topology on $\mathscr{G}^{\mathrm{hol},\mathbb{R}}_{\mathsf{M},\overline{\mathsf{E}}}$ described above, along with the preceding lemma, to immediately give a locally convex topology for $\Gamma^{\omega}(\mathsf{E})$ that we refer to as the ***direct C^ω-topology***.

Let us make an important observation about the direct C^{ω}-topology. Let us denote by $\mathscr{S}_{\mathsf{M}}$ the set of all Stein neighbourhoods of M in $\overline{\mathsf{M}}$. As shown by Grauert [11, Sect. 3.4], if $\overline{\mathcal{U}} \in \mathscr{N}_{\mathsf{M}}$, then there exists $\overline{\mathcal{S}} \in \mathscr{S}_{\mathsf{M}}$ with $\overline{\mathcal{S}} \subseteq \overline{\mathcal{U}}$. Therefore, $\mathscr{S}_{\mathsf{M}}$ is cofinal in $\mathscr{N}_{\mathsf{M}}$ and so the directed systems $(\Gamma^{\mathrm{hol}}(\overline{\mathsf{E}}|\overline{\mathcal{U}}))_{\overline{\mathcal{U}} \in \mathscr{N}_{\mathsf{M}}}$ and $(\Gamma^{\mathrm{hol}}(\overline{\mathsf{E}}|\overline{\mathcal{S}}))_{\overline{\mathcal{S}} \in \mathscr{S}_{\mathsf{M}}}$ induce the same final topology on $\Gamma^{\omega}(\mathsf{E})$ [12, p. 137].

5.2 Topologies for Germs of Holomorphic Functions About Compact Sets

In the preceding section, we gave a more or less direct description of a topology for the space of real analytic sections. This description has a benefit of being the one that one might naturally arrive at after some thought. However, there is not a lot that one can do with this description of the topology, e.g., many of the useful properties of the space of real analytic sections cannot be ascertained from this description. In this section we develop the means by which one can consider alternative descriptions of this topology that, for example, lead to explicit seminorms for the topology on the space of real analytic sections. These seminorms will be an essential part of our developing a useful theory for time-varying real analytic vector fields.

5.2.1 The Direct Limit Topology for the Space of Germs About a Compact Set

We continue with the notation from Sect. 5.1.2. For $K \subseteq \mathsf{M}$ compact, we have the direct limit topology, described above for general subsets $A \subseteq \mathsf{M}$, on $\mathscr{G}^{\mathrm{hol},\mathbb{R}}_{K,\overline{\mathsf{E}}}$. We seem to have gained nothing, since we have yet another direct limit topology. However, the direct limit can be shown to be of a friendly sort as follows. Unlike the general situation, since K is compact there is a *countable* family $(\overline{\mathcal{U}}_{K,j})_{j\in\mathbb{Z}_{>0}}$ of relatively compact neighbourhoods from $\mathscr{N}_K$ with the property that $\mathrm{cl}(\overline{\mathcal{U}}_{K,j+1}) \subseteq \overline{\mathcal{U}}_{K,j}$ and $K = \cap_{j\in\mathbb{Z}_{>0}} \overline{\mathcal{U}}_{K,j}$. Moreover, the sequence $(\overline{\mathcal{U}}_{K,j})_{j\in\mathbb{Z}_{>0}}$ is cofinal in $\mathscr{N}_K$, i.e., if $\overline{\mathcal{U}} \in \mathscr{N}_K$, then there exists $j \in \mathbb{Z}_{>0}$ with $\overline{\mathcal{U}}_{K,j} \subseteq \overline{\mathcal{U}}$. Let us fix such a family of neighbourhoods. Let us fix $j \in \mathbb{Z}_{>0}$ for a moment. Let $\Gamma^{\mathrm{hol},\mathbb{R}}_{\mathrm{bdd}}(\overline{\mathsf{E}}|\overline{\mathcal{U}}_{K,j})$ be the set of bounded sections from $\Gamma^{\mathrm{hol},\mathbb{R}}(\overline{\mathsf{E}}|\overline{\mathcal{U}}_{K,j})$, boundedness being taken relative to the Hermitian fibre metric $\overline{\mathbb{G}}_0$. As we have seen in Lemma 4.1, if we define a norm on $\Gamma^{\mathrm{hol},\mathbb{R}}_{\mathrm{bdd}}(\overline{\mathsf{E}}|\overline{\mathcal{U}}_{K,j})$ by

$$p^{\mathrm{hol}}_{\overline{\mathcal{U}}_{K,j},\infty}(\overline{\xi}) = \sup\{\|\overline{\xi}(\overline{x})\|_{\overline{\mathbb{G}}_0} \mid \overline{x} \in \overline{\mathcal{U}}_{K,j}\},$$

then this makes $\Gamma^{\mathrm{hol},\mathbb{R}}_{\mathrm{bdd}}(\overline{\mathcal{U}}_{K,j})$ into a Banach space, a closed subspace of the Banach space of bounded continuous sections of $\overline{\mathsf{E}}|\overline{\mathcal{U}}_{K,j}$. Now, no longer fixing j, we have a sequence of inclusions

$$\Gamma^{\mathrm{hol},\mathbb{R}}_{\mathrm{bdd}}(\overline{\mathsf{E}}|\overline{\mathcal{U}}_{K,1}) \subseteq \Gamma^{\mathrm{hol},\mathbb{R}}(\overline{\mathsf{E}}|\overline{\mathcal{U}}_{K,1}) \subseteq \Gamma^{\mathrm{hol},\mathbb{R}}_{\mathrm{bdd}}(\overline{\mathsf{E}}|\overline{\mathcal{U}}_{K,2}) \subseteq \cdots \subseteq \Gamma^{\mathrm{hol},\mathbb{R}}(\overline{\mathsf{E}}|\overline{\mathcal{U}}_{K,j}) \subseteq \Gamma^{\mathrm{hol},\mathbb{R}}_{\mathrm{bdd}}(\overline{\mathsf{E}}|\overline{\mathcal{U}}_{K,j+1}) \subseteq \cdots.$$

The inclusion $\Gamma^{\mathrm{hol},\mathbb{R}}(\overline{\mathcal{U}}_{K,j}) \subseteq \Gamma^{\mathrm{hol},\mathbb{R}}_{\mathrm{bdd}}(\overline{\mathcal{U}}_{K,j+1})$, $j \in \mathbb{Z}_{>0}$, is by restriction from $\overline{\mathcal{U}}_{K,j}$ to the smaller $\overline{\mathcal{U}}_{K,j+1}$, keeping in mind that $\mathrm{cl}(\overline{\mathcal{U}}_{K,j+1}) \subseteq \overline{\mathcal{U}}_{K,j}$. By Lemma 4.1, all inclusions are continuous. For $j \in \mathbb{Z}_{>0}$ define

$$\begin{aligned} r_{K,j}\colon \Gamma^{\mathrm{hol},\mathbb{R}}_{\mathrm{bdd}}(\overline{\mathsf{E}}|\overline{\mathcal{U}}_{K,j}) &\to \mathscr{G}^{\mathrm{hol},\mathbb{R}}_{K,\overline{\mathsf{E}}} \\ \overline{\xi} &\mapsto [\overline{\xi}]_K. \end{aligned} \tag{5.1}$$

Now one can show that the direct limit topologies induced on $\mathscr{G}^{\mathrm{hol},\mathbb{R}}_{K,\overline{\mathsf{E}}}$ by the directed system $(\Gamma^{\mathrm{hol},\mathbb{R}}(\overline{\mathsf{E}}|\overline{\mathcal{U}}))_{\overline{\mathcal{U}}\in\mathscr{N}_K}$ of Fréchet spaces and by the directed system $(\Gamma^{\mathrm{hol},\mathbb{R}}_{\mathrm{bdd}}(\overline{\mathsf{E}}|\overline{\mathcal{U}}_{K,j}))_{j\in\mathbb{Z}_{>0}}$ of Banach spaces agree [15, Theorem 8.4]. We refer to [3], starting on page 63, for a fairly comprehensive discussion of the topology we have just described in the context of germs of holomorphic functions about a compact subset $K \subseteq \mathbb{C}^n$.

5.2.2 A Weighted Direct Limit Topology for Sections of Bundles of Infinite Jets

Here we provide a direct limit topology for a subspace of the space of continuous sections of the infinite jet bundle of a vector bundle. Below we shall connect this direct limit topology to the direct limit topology described above for germs of holomorphic sections about a compact set. The topology we give here has the advantage of providing explicit seminorms for the topology of germs, and subsequently for the space of real analytic sections.

For this description, we work with infinite jets, so let us introduce the notation we will use for this, referring to [21, Chap. 7] for details. Let us denote by $\mathsf{J}^\infty\mathsf{E}$ the bundle of infinite jets of a vector bundle $\pi\colon \mathsf{E} \to \mathsf{M}$, this being the inverse limit (in the category of sets, for the moment) of the inverse system $(\mathsf{J}^m\mathsf{E})_{m\in\mathbb{Z}_{\geq 0}}$ with mappings π_m^{m+1}, $m \in \mathbb{Z}_{\geq 0}$. Precisely,

$$\mathsf{J}^\infty\mathsf{E} = \Big\{\phi \in \prod_{m\in\mathbb{Z}_{\geq 0}} \mathsf{J}^m\mathsf{E} \;\Big|\; \pi_l^k \circ \phi(k) = \phi(l),\ k, l \in \mathbb{Z}_{\geq 0},\ k \geq l\Big\}.$$

We let $\pi_m^\infty\colon \mathsf{J}^\infty\mathsf{E} \to \mathsf{J}^m\mathsf{E}$ be the projection defined by $\pi_m^\infty(\phi) = \phi(m)$. For $\xi \in \Gamma^\infty(\mathsf{E})$ we let $j_\infty\xi\colon \mathsf{M} \to \mathsf{J}^\infty\mathsf{E}$ be defined by $\pi_m^\infty \circ j_\infty\xi(x) = j_m\xi(x)$. By a theorem of Borel [5], if $\phi \in \mathsf{J}^\infty\mathsf{E}$, there exist $\xi \in \Gamma^\infty(\mathsf{E})$ and $x \in \mathsf{M}$ such that $j_\infty\xi(x) = \phi$. We can define sections of $\mathsf{J}^\infty\mathsf{E}$ in the usual manner: a section is a map $\Xi\colon \mathsf{M} \to \mathsf{J}^\infty\mathsf{E}$ satisfying $\pi_0^\infty \circ \Xi(x) = x$ for every $x \in \mathsf{M}$. We shall equip $\mathsf{J}^\infty\mathsf{E}$ with the initial topology so that a section Ξ is continuous if and only if $\pi_m^\infty \circ \Xi$ is continuous for every $m \in \mathbb{Z}_{\geq 0}$. We denote the space of continuous sections of $\mathsf{J}^\infty\mathsf{E}$ by $\Gamma^0(\mathsf{J}^\infty\mathsf{E})$. Since we are only dealing with continuous sections, we can talk about sections defined on any subset $A \subseteq \mathsf{M}$, using the relative topology on A. The continuous sections defined on $A \subseteq \mathsf{M}$ will be denoted by $\Gamma^0(\mathsf{J}^\infty\mathsf{E}|A)$.

Now let $K \subseteq \mathsf{M}$ be compact and, for $j \in \mathbb{Z}_{>0}$, denote

$$\mathscr{E}_j(K) = \{\Xi \in \Gamma^0(\mathsf{J}^\infty\mathsf{E}|K) \mid \sup\{j^{-m}\|\pi_m^\infty \circ \Xi(x)\|_{\overline{\mathbb{G}}_m} \mid m \in \mathbb{Z}_{\geq 0},\ x \in K\} < \infty\},$$

and on $\mathscr{E}_j(K)$ we define a norm $p_{K,j}$ by

$$p_{K,j}(\Xi) = \sup\{j^{-m}\|\pi_m^\infty \circ \Xi(x)\|_{\overline{\mathbb{G}}_m} \mid m \in \mathbb{Z}_{\geq 0},\ x \in K\}.$$

One readily verifies that, for each $j \in \mathbb{Z}_{>0}$, $(\mathscr{E}_j(K), p_{K,j})$ is a Banach space. Note that $\mathscr{E}_j(K) \subseteq \mathscr{E}_{j+1}(K)$ and that $p_{K,j+1}(\Xi) \leq p_{K,j}(\Xi)$ for $\Xi \in \mathscr{E}_j(K)$, and so the inclusion of $\mathscr{E}_j(K)$ in $\mathscr{E}_{j+1}(K)$ is continuous. We let $\mathscr{E}(K)$ be the direct limit of the directed system $(\mathscr{E}_j(K))_{j\in\mathbb{Z}_{>0}}$.

We shall subsequently explore more closely the relationship between the direct limit topology for $\mathscr{E}(K)$ and the topology for $\mathscr{G}^{\mathrm{hol},\mathbb{R}}_{K,\overline{\mathsf{E}}}$. For now, we merely observe that the direct limit topology for $\mathscr{E}(K)$ admits a characterisation by seminorms. To state the result, let us denote by $c_0(\mathbb{Z}_{\geq 0}; \mathbb{R}_{>0})$ the set of sequences $(a_m)_{m\in\mathbb{Z}_{\geq 0}}$ in $\mathbb{R}_{>0}$ that converge to 0. Let us abbreviate such a sequence by $\boldsymbol{a} = (a_m)_{m\in\mathbb{Z}_{\geq 0}}$. The following result is modelled after [24, Lemma 1].

Lemma 5.3 (Seminorms for $\mathscr{E}(K)$). *The direct limit topology for $\mathscr{E}(K)$ is defined by the seminorms*

$$p_{K,\boldsymbol{a}} = \sup\{a_0 a_1 \cdots a_m \|\pi_m^\infty \circ \Xi(x)\|_{\overline{\mathbb{G}}_m} \mid m \in \mathbb{Z}_{\geq 0},\ x \in K\},$$

for $\boldsymbol{a} \in \mathrm{c}_0(\mathbb{Z}_{\geq 0}; \mathbb{R}_{>0})$.

Proof: First we show that the seminorms $p_{K,\boldsymbol{a}}$, $\boldsymbol{a} \in \mathrm{c}_0(\mathbb{Z}_{\geq 0}; \mathbb{R}_{>0})$, are continuous on $\mathscr{E}(K)$. It suffices to show that $p_{K,\boldsymbol{a}}|\mathscr{E}_j(K)$ is continuous for each $j \in \mathbb{Z}_{>0}$ [7, Proposition IV.5.7]. Thus, since $\mathscr{E}_j(K)$ is a Banach space, it suffices to show that, if $(\Xi_k)_{k\in\mathbb{Z}_{>0}}$ is a sequence in $\mathscr{E}_j(K)$ converging to zero, then $\lim_{k\to\infty} p_{K,\boldsymbol{a}}(\Xi_k) = 0$. Let $N \in \mathbb{Z}_{\geq 0}$ be such that $a_N < \frac{1}{j}$. Let $C \geq 1$ be such that

$$a_0 a_1 \cdots a_m \leq C j^{-m}, \qquad m \in \{0, 1, \dots, N\},$$

this being possible since there are only finitely many inequalities to satisfy. Therefore, for any $m \in \mathbb{Z}_{\geq 0}$, we have $a_0 a_1 \cdots a_m \leq C j^{-m}$. Then, for any $\Xi \in \Gamma^0(\mathsf{J}^\infty\mathsf{E}|K)$,

$$a_0 a_1 \cdots a_m \|\pi_m^\infty \circ \Xi(x)\|_{\overline{\mathbb{G}}_m} \leq C j^{-m} \|\pi_m^\infty \circ \Xi(x)\|_{\overline{\mathbb{G}}_m}$$

for every $x \in K$ and $m \in \mathbb{Z}_{\geq 0}$. From this we immediately have $\lim_{k\to\infty} p_{K,\boldsymbol{a}}(\Xi_k) = 0$, as desired. This shows that the direct limit topology on $\mathscr{E}(K)$ is stronger than the topology defined by the family of seminorms $p_{K,\boldsymbol{a}}$, $\boldsymbol{a} \in \mathrm{c}_0(\mathbb{Z}_{\geq 0}; \mathbb{R}_{>0})$.

For the converse, we show that every neighbourhood of $0 \in \mathscr{E}(K)$ in the direct limit topology contains a neighbourhood of zero in the topology defined by the seminorms $p_{K,\boldsymbol{a}}$, $\boldsymbol{a} \in \mathrm{c}_0(\mathbb{Z}_{\geq 0}; \mathbb{R}_{>0})$. Let $\mathcal{B}_j$ denote the unit ball in $\mathscr{E}_j(K)$. A neighbourhood of 0 in the direct limit topology contains a union of balls $\epsilon_j \mathcal{B}_j$ for some $\epsilon_j \in \mathbb{R}_{>0}$, $j \in \mathbb{Z}_{>0}$, (see [22, p. 54]) and we can assume, without loss of generality, that $\epsilon_j \in (0, 1)$ for each $j \in \mathbb{Z}_{>0}$. We define an increasing sequence $(m_j)_{j\in\mathbb{Z}_{>0}}$ in $\mathbb{Z}_{\geq 0}$ as follows. Let $m_1 = 0$. Having defined $m_1, \dots, m_j$, define $m_{j+1} > m_j$ by requiring that $j < \epsilon_{j+1}^{1/m_{j+1}}(j+1)$. For $m \in \{m_j, \dots, m_{j+1} - 1\}$, define $a_m \in \mathbb{R}_{>0}$ by $a_m^{-1} = \epsilon_j^{1/m_j} j$. Note that, for $m \in \{m_j, \dots, m_{j+1} - 1\}$, we have

$$a_m^{-m} = \epsilon_j^{m/m_j} j^m \leq \epsilon_j j^m.$$

Note that $\lim_{m\to\infty} a_m = 0$. If $\Xi \in \Gamma^0(\mathsf{J}^\infty\mathsf{E}|K)$ satisfies $p_{K,\boldsymbol{a}}(\Xi) \leq 1$, then, for $m \in \{m_j, \dots, m_{j+1} - 1\}$, we have

$$\begin{aligned} j^{-m} \|\pi_m^\infty \circ \Xi(x)\|_{\overline{\mathbb{G}}_m} &\leq a_m^m \epsilon_j \|\pi_m^\infty \circ \Xi(x)\|_{\overline{\mathbb{G}}_m} \\ &\leq a_0 a_1 \cdots a_m \epsilon_j \|\pi_m^\infty \circ \Xi(x)\|_{\overline{\mathbb{G}}_m} \leq \epsilon_j \end{aligned}$$

for $x \in K$. Thus, if $\Xi \in \Gamma^0(\mathsf{J}^\infty\mathsf{E}|K)$ satisfies $p_{K,\boldsymbol{a}}(\Xi) \leq 1$, then, for $m \in \{m_j, \dots, m_{j+1} - 1\}$, we have $\pi_m^\infty \circ \Xi \in \epsilon_j \mathcal{B}_j$. Therefore, $\Xi \in \cup_{j\in\mathbb{Z}_{>0}} \epsilon_j \mathcal{B}_j$, and this shows that, for $\boldsymbol{a}$ as constructed above,

$$\{\Xi \in \Gamma^0(\mathsf{J}^\infty\mathsf{E}|K) \mid p_{K,\boldsymbol{a}}(\Xi) \leq 1\} \subseteq \cup_{j\in\mathbb{Z}_{>0}} \epsilon_j \mathcal{B}_j,$$

giving the desired conclusion. □

The following attribute of the direct limit topology for $\mathscr{E}(K)$ will also be useful.

Lemma 5.4 (**$\mathscr{E}(K)$ is a Regular Direct Limit**). *The direct limit topology for $\mathscr{E}(K)$ is regular, i.e., if $\mathcal{B} \subseteq \mathscr{E}(K)$ is von Neumann bounded, then there exists $j \in \mathbb{Z}_{>0}$ such that $\mathcal{B}$ is contained in and von Neumann bounded in $\mathscr{E}_j(K)$.*

Proof: Let $\mathcal{B}_j \subseteq \mathscr{E}_j(K)$, $j \in \mathbb{Z}_{>0}$, be the closed unit ball with respect to the norm topology. We claim that $\mathcal{B}_j$ is closed in the direct limit topology of $\mathscr{E}(K)$. To prove this, we shall prove that $\mathcal{B}_j$ is closed in a topology that is weaker than the direct limit topology.

The weaker topology we use is the topology induced by the topology of pointwise convergence in $\Gamma^0(\mathsf{J}^\infty\mathsf{E}|K)$. To be precise, let $\mathscr{E}'_j(K)$ be the vector space $\mathscr{E}_j(K)$ with the topology defined by the seminorms

$$p_{x,j}(\Xi) = \sup\{j^{-m}\|\pi^\infty_m \circ \Xi(x)\|_{\overline{\mathbb{G}}_m} \mid m \in \mathbb{Z}_{\geq 0}\}, \qquad x \in K.$$

Clearly the identity map from $\mathscr{E}_j(K)$ to $\mathscr{E}'_j(K)$ is continuous, and so the topology of $\mathscr{E}'_j(K)$ is weaker than the usual topology of $\mathscr{E}(K)$. Now let $\mathscr{E}'(K)$ be the direct limit of the directed system $(\mathscr{E}'_j(K))_{j\in\mathbb{Z}_{>0}}$. Note that, algebraically, $\mathscr{E}'(K) = \mathscr{E}(K)$, but the spaces have different topologies, the topology for $\mathscr{E}'(K)$ being weaker than that for $\mathscr{E}(K)$.

We will show that $\mathcal{B}_j$ is closed in $\mathscr{E}'(K)$. Let $(I, \preceq)$ be a directed set and let $(\Xi_i)_{i\in I}$ be a convergent net in $\mathcal{B}_j$ in the topology of $\mathscr{E}'(K)$. Thus we have a map $\Xi\colon K \to \mathsf{J}^\infty\mathsf{E}|K$ such that, for each $x \in K$, $\lim_{i\in I} \Xi_i(x) = \Xi(x)$. If $\Xi \notin \mathcal{B}_j$, then there exists $x \in K$ such that

$$\sup\{j^{-m}\|\pi^\infty_m \circ \Xi(x)\|_{\overline{\mathbb{G}}_m} \mid m \in \mathbb{Z}_{\geq 0}\} > 1.$$

Let $\epsilon \in \mathbb{R}_{>0}$ be such that

$$\sup\{j^{-m}\|\pi^\infty_m \circ \Xi(x)\|_{\overline{\mathbb{G}}_m} \mid m \in \mathbb{Z}_{\geq 0}\} > 1 + \epsilon$$

and let $i_0 \in I$ be such that

$$\sup\{j^{-m}\|\pi^\infty \circ \Xi_i(x) - \pi^\infty_m \circ \Xi(x)\|_{\overline{\mathbb{G}}_m} \mid m \in \mathbb{Z}_{\geq 0}\} < \epsilon$$

for $i_0 \preceq i$, this by pointwise convergence. We thus have, for all $i_0 \preceq i$,

$$\begin{aligned}\epsilon &< \sup\{j^{-m}\|\pi^\infty_m \circ \Xi(x)\|_{\overline{\mathbb{G}}_m} \mid m \in \mathbb{Z}_{\geq 0}\} - \sup\{j^{-m}\|\pi^\infty_m \circ \Xi_i(x)\|_{\overline{\mathbb{G}}_m} \mid m \in \mathbb{Z}_{\geq 0}\} \\ &\leq \sup\{j^{-m}\|\pi^\infty_m \circ \Xi_i(x) - \pi^\infty_m \circ \Xi(x)\|_{\overline{\mathbb{G}}_m} \mid m \in \mathbb{Z}_{\geq 0}\} < \epsilon,\end{aligned}$$

which contradiction gives the conclusion that $\Xi \in \mathcal{B}_j$.

Since $\mathcal{B}_j$ has been shown to be closed in $\mathscr{E}(K)$, the lemma now follows from [3, Corollary 7]. ◻

5.2.3 Seminorms for the Topology of Spaces of Holomorphic Germs

Let us define seminorms $p^\omega_{K,\boldsymbol{a}}$, $K \subseteq \mathsf{M}$ compact, $\boldsymbol{a} \in c_0(\mathbb{Z}_{\geq 0}; \mathbb{R}_{>0})$, for $\mathscr{G}^{\text{hol},\mathbb{R}}_{K,\overline{\mathsf{E}}}$ by

$$p^\omega_{K,\boldsymbol{a}}([\overline{\xi}]_K) = \sup\{a_0 a_1 \cdots a_m \| j_m \overline{\xi}(x) \|_{\overline{\mathbb{G}}_m} \mid x \in K,\ m \in \mathbb{Z}_{\geq 0}\}.$$

We can (and will) also think of $p^\omega_{K,\boldsymbol{a}}$ as being a seminorm on $\Gamma^\omega(\mathsf{E})$ defined by the same formula.

Let us prove that the seminorms $p^\omega_{K,\boldsymbol{a}}$, $K \subseteq \mathsf{M}$ compact, $\boldsymbol{a} \in c_0(\mathbb{Z}_{\geq 0}; \mathbb{R}_{>0})$, can be used to define the direct limit topology on $\mathscr{G}^{\text{hol},\mathbb{R}}_{K,\overline{\mathsf{E}}}$.

Theorem 5.5 (Seminorms for $\mathscr{G}^{\text{hol},\mathbb{R}}_{K,\overline{\mathsf{E}}}$). *Let $\pi\colon \mathsf{E} \to \mathsf{M}$ be a real analytic vector bundle and let $K \subseteq \mathsf{M}$ be compact. Then the family of seminorms $p^\omega_{K,\boldsymbol{a}}$, $\boldsymbol{a} \in c_0(\mathbb{Z}_{\geq 0}; \mathbb{R}_{>0})$, defines a locally convex topology on $\mathscr{G}^{\text{hol},\mathbb{R}}_{K,\overline{\mathsf{E}}}$ agreeing with the direct limit topology.*

Proof: Let us sketch the idea of the proof before we begin. The essential construction is a natural mapping L_K from $\mathscr{G}^{\text{hol},\mathbb{R}}_{K,\overline{\mathsf{E}}}$ to $\mathscr{E}(K)$ corresponding to the fact that an holomorphic section is determined by its infinite jet. By some not completely obvious, but straightforward, arguments we show that L_K is an homeomorphism onto its image. This is proved in Lemma 1. The most difficult part of the proof is to show that preimages of bounded sets are bounded. It is here that we make essential use of the difficult Lemma 2.5, along with more or less elementary facts about power series. After we show that L_K is an homeomorphism onto its image, the theorem follows easily from Lemma 5.3 since this lemma gives seminorms for $\mathscr{E}(K)$.

Let $K \subseteq \mathsf{M}$ be compact and let $(\overline{\mathcal{U}}_j)_{j \in \mathbb{Z}_{>0}}$ be a sequence of neighbourhoods of K in $\overline{\mathsf{M}}$ such that $\mathrm{cl}(\overline{\mathcal{U}}_{j+1}) \subseteq \overline{\mathcal{U}}_j$, $j \in \mathbb{Z}_{>0}$, and such that $K = \cap_{j \in \mathbb{Z}_{>0}} \overline{\mathcal{U}}_j$. We have mappings

$$\begin{aligned} r_{\overline{\mathcal{U}}_j,K}\colon \Gamma^{\text{hol},\mathbb{R}}_{\text{bdd}}(\overline{\mathsf{E}}|\overline{\mathcal{U}}_j) &\to \mathscr{G}^{\text{hol},\mathbb{R}}_{K,\overline{\mathsf{E}}} \\ \overline{\xi} &\mapsto [\overline{\xi}]_K. \end{aligned}$$

The maps $r_{\overline{\mathcal{U}}_j,K}$ can be assumed to be injective without loss of generality, by making sure that each open set $\overline{\mathcal{U}}_j$ consists of disconnected neighbourhoods of the connected components of K. Since M is Hausdorff and the connected components of K are compact, this can always be done by choosing the initial open set $\overline{\mathcal{U}}_1$ sufficiently small. In this way, $\Gamma^{\text{hol},\mathbb{R}}_{\text{bdd}}(\overline{\mathsf{E}}|\overline{\mathcal{U}}_j)$, $j \in \mathbb{Z}_{>0}$, are regarded as subspaces of $\mathscr{G}^{\text{hol},\mathbb{R}}_{K,\overline{\mathsf{E}}}$. It is convenient to be able to do this.

We will work with the locally convex space $\mathscr{E}(K)$ introduced in Sect. 5.2.2 and define a mapping $L_K\colon \mathscr{G}^{\text{hol},\mathbb{R}}_{K,\overline{\mathsf{E}}} \to \mathscr{E}(K)$ by $L_K([\overline{\xi}]_K) = j_\infty \xi | K$. Let us prove that this mapping is well defined, i.e., show that, if $[\overline{\xi}]_K \in \mathscr{G}^{\text{hol},\mathbb{R}}_{K,\overline{\mathsf{E}}}$, then $L_K([\overline{\xi}]_K) \in \mathscr{E}_j(K)$ for some $j \in \mathbb{Z}_{>0}$. Let $\overline{\mathcal{U}}$ be a neighbourhood of K in $\overline{\mathsf{M}}$ on which the section $\overline{\xi}$ is defined, holomorphic, and bounded. Then $\xi|(\mathsf{M} \cap \overline{\mathcal{U}})$ is real analytic and so, by Lemma 2.6, there exist $C, r \in \mathbb{R}_{>0}$ such that

$$\|j_m\xi(x)\|_{\overline{\mathbb{G}}_m} \leq Cr^{-m}, \qquad x \in K,\ m \in \mathbb{Z}_{\geq 0}.$$

If $j > r^{-1}$ it immediately follows that

$$\sup\{j^{-m}\|j_m\xi(x)\|_{\overline{\mathbb{G}}_m} \mid x \in K,\ m \in \mathbb{Z}_{\geq 0}\} < \infty,$$

i.e., $L_K([\overline{\xi}]_K) \in \mathscr{E}_j(K)$.

The following lemma records the essential feature of L_K.

Lemma 1 *The mapping L_K is a continuous, injective, open mapping, and so an homeomorphism onto its image.*

Proof: To show that L_K is continuous, it suffices to show that $L_K|\Gamma^{\mathrm{hol},\mathbb{R}}_{\mathrm{bdd}}(\overline{\mathsf{E}}|\overline{\mathcal{U}}_j)$ is continuous for each $j \in \mathbb{Z}_{\geq 0}$. We will show this by showing that, for each $j \in \mathbb{Z}_{>0}$, there exists $j' \in \mathbb{Z}_{>0}$ such that $L_K(\Gamma^{\mathrm{hol}}_{\mathrm{bdd}}(\overline{\mathsf{E}}|\overline{\mathcal{U}}_j)) \subseteq \mathscr{E}_{j'}(K)$ and such that L_K is continuous as a map from $\Gamma^{\mathrm{hol}}_{\mathrm{bdd}}(\overline{\mathsf{E}}|\overline{\mathcal{U}}_j)$ to $\mathscr{E}_{j'}(K)$. Since $\mathscr{E}_{j'}(K)$ is continuously included in $\mathscr{E}(K)$, this will give the continuity of L_K. First let us show that $L_K(\Gamma^{\mathrm{hol}}_{\mathrm{bdd}}(\overline{\mathsf{E}}|\overline{\mathcal{U}}_j)) \subseteq \mathscr{E}_{j'}(K)$ for some $j' \in \mathbb{Z}_{>0}$. By Proposition 4.2, there exist $C, r \in \mathbb{R}_{>0}$ such that

$$\|j_m\xi(x)\|_{\overline{\mathbb{G}}_m} \leq Cr^{-m}p^{\mathrm{hol}}_{\overline{\mathcal{U}}_j,\infty}(\overline{\xi})$$

for every $m \in \mathbb{Z}_{\geq 0}$ and $\overline{\xi} \in \Gamma^{\mathrm{hol}}_{\mathrm{bdd}}(\overline{\mathsf{E}}|\overline{\mathcal{U}}_j)$. Taking $j' \in \mathbb{Z}_{>0}$ such that $j' \geq r^{-1}$ we have $L_K(\Gamma^{\mathrm{hol}}_{\mathrm{bdd}}(\overline{\mathsf{E}}|\overline{\mathcal{U}}_j)) \subseteq \mathscr{E}_{j'}(K)$, as claimed. To show that L_K is continuous as a map from $\Gamma^{\mathrm{hol}}_{\mathrm{bdd}}(\overline{\mathsf{E}}|\overline{\mathcal{U}}_j)$ to $\mathscr{E}_{j'}(K)$, let $([\xi_k]_K)_{k\in\mathbb{Z}_{>0}}$ be a sequence in $\Gamma^{\mathrm{hol}}_{\mathrm{bdd}}(\overline{\mathsf{E}}|\overline{\mathcal{U}}_j)$ converging to zero. We then have

$$\lim_{k\to\infty} \sup\{(j')^{-m}\|j_m\xi_k(x)\|_{\overline{\mathbb{G}}} \mid x \in K,\ m \in \mathbb{Z}_{\geq 0}\}$$
$$\leq \lim_{k\to\infty} C\sup\{\|\overline{\xi}_k(z)\|_{\overline{\mathbb{G}}} \mid z \in \overline{\mathcal{U}}_j\} = 0,$$

giving the desired continuity.

Since germs of holomorphic sections are uniquely determined by their infinite jets, injectivity of L_K follows.

We claim that, if $\mathcal{B} \subseteq \mathscr{E}(K)$ is von Neumann bounded, then $L_K^{-1}(\mathcal{B})$ is also von Neumann bounded. By Lemma 5.4, if $\mathcal{B} \subseteq \mathscr{E}(K)$ is bounded, then $\mathcal{B}$ is contained and bounded in $\mathscr{E}_j(K)$ for some $j \in \mathbb{Z}_{>0}$. Therefore, there exists $C \in \mathbb{R}_{>0}$ such that, if $L_K([\overline{\xi}]_K) \subseteq \mathcal{B}$, then

$$\|j_m\xi(x)\|_{\overline{\mathbb{G}}_m} \leq Cj^{-m}, \qquad x \in K,\ m \in \mathbb{Z}_{\geq 0}.$$

Let $x \in K$ and let $(\mathcal{V}_x, \psi_x)$ be a vector bundle chart for E about x with corresponding chart $(\mathcal{U}_x, \phi_x)$ for M. Suppose the fibre dimension of E over $\mathcal{U}_x$ is k and that ϕ_x takes values in $\mathbb{R}^n$. Let $\mathcal{U}'_x \subseteq \mathcal{U}_x$ be a relatively compact neighbourhood of x such that

$\mathrm{cl}(\mathcal{U}'_x) \subseteq \mathcal{U}_x$. Denote $K_x = K \cap \mathrm{cl}(\mathcal{U}'_x)$. By Lemma 2.5, there exist $C_x, r_x \in \mathbb{R}_{>0}$ such that, if $L_K([\overline{\xi}]_K) \subseteq \mathcal{B}$, then

$$|\boldsymbol{D}^I \xi^a(\boldsymbol{x})| \leq C_x I! r_x^{-|I|}, \qquad \boldsymbol{x} \in \phi_x(K_x),\ I \in \mathbb{Z}^n_{\geq 0},\ a \in \{1, \dots, k\},$$

where $\boldsymbol{\xi}$ is the local representative of ξ. Note that this implies the following for each $[\overline{\xi}]_K$ such that $L_K([\overline{\xi}]_K) \subseteq \mathcal{B}$ and for each $a \in \{1, \dots, k\}$:

1. $\overline{\xi}^a$ admits a convergent power series expansion to an holomorphic function on the polydisk $\mathsf{D}(\boldsymbol{\sigma}_x, \phi_x(x))$ for $\sigma_x < r_x$;
2. on the polydisk $\mathsf{D}(\boldsymbol{\sigma}_x, \phi_x(x))$, $\overline{\xi}^a$ satisfies $|\overline{\xi}^a| \leq (\frac{1}{1-\sigma_x})^n$.

It follows that, if $L_K([\overline{\xi}]_K) \in \mathcal{B}$, then $\overline{\xi}$ has a bounded holomorphic extension in some coordinate polydisk around each $x \in K$. By a standard compactness argument and since $\cap_{j \in \mathbb{Z}_{>0}} \overline{\mathcal{U}}_j = K$, there exists $j' \in \mathbb{Z}_{>0}$ such that $\overline{\xi} \in \Gamma^{\mathrm{hol},\mathbb{R}}_{\mathrm{bdd}}(\overline{\mathsf{E}}|\overline{\mathcal{U}}_{j'})$ for each $[\overline{\xi}]_K$ such that $L_K([\overline{\xi}]_K) \in \mathcal{B}$, and that the set of such sections of $\overline{\mathsf{E}}|\overline{\mathcal{U}}_{j'}$ is von Neumann bounded, i.e., norm bounded. Thus $L_K^{-1}(\mathcal{B})$ is von Neumann bounded, as claimed.

Note also that $\mathscr{E}(K)$ is a DF-space since Banach spaces are DF-spaces [14, Corollary 12.4.4] and countable direct limits of DF-spaces are DF-spaces [14, Theorem 12.4.8]. Therefore, by the open mapping lemma from Sect. 2 from [2], the result follows. ▽

From the lemma, it follows that the direct limit topology of $\mathscr{G}^{\mathrm{hol},\mathbb{R}}_{K,\overline{\mathsf{E}}}$ agrees with that induced by its image in $\mathscr{E}(K)$. Since the seminorms $p_{K,\boldsymbol{a}}$, $\boldsymbol{a} \in c_0(\mathbb{Z}_{\geq 0}; \mathbb{R}_{>0})$, define the locally convex topology of $\mathscr{E}(K)$ by Lemma 5.3, it follows that the seminorms $p^{\omega}_{K,\boldsymbol{a}}$, $\boldsymbol{a} \in c_0(\mathbb{Z}_{\geq 0}; \mathbb{R}_{>0})$, define the direct limit topology of $\mathscr{G}^{\mathrm{hol},\mathbb{R}}_{K,\overline{\mathsf{E}}}$. □

The problem of providing seminorms for the direct limit topology of $\mathscr{G}^{\mathrm{hol},\mathbb{R}}_{K,\overline{\mathsf{E}}}$ is a nontrivial one, so let us provide a little history for what led to the preceding theorem. First of all, the first concrete characterisation of seminorms for germs of holomorphic functions about compact subsets of $\mathbb{C}^n$ comes in [18]. Mujica [18] provides seminorms having two parts, one very much resembling the seminorms we use, and another part that is more complicated. These seminorms specialise to the case where the compact set lies in $\mathbb{R}^n \subseteq \mathbb{C}^n$, and the first mention of this we have seen in the research literature is in the notes of Domanski [8]. The first full proof that the seminorms analogous to those we define are, in fact, the seminorms for the space of real analytic functions on open subsets of $\mathbb{R}^n$ appears in the recent note of Vogt [24]. Our presentation is an adaptation, not quite trivial as it turns out, of Vogt's constructions. One of the principal difficulties is Lemma 2.5 which is essential in showing that our jet bundle fibre metrics $\|\cdot\|_{\overline{\mathbb{G}}_m}$ are suitable for defining the seminorms for the real analytic topology. Note that one cannot use arbitrary fibre metrics, since one needs to have the behaviour of these metrics be regulated to the real analytic topology as the order of jets goes to infinity. Because our fibre metrics are constructed by differentiating objects defined at low order, i.e., the connections ∇ and ∇^0, we can ensure that the fibre metrics are compatible with real analytic growth conditions on derivatives.

5.2.4 An Inverse Limit Topology for the Space of Real Analytic Sections

In the preceding three sections we provided three topologies for the space $\mathscr{G}^{\mathrm{hol},\mathbb{R}}_{K,\overline{\mathsf{E}}}$ of holomorphic sections about a compact subset K of a real analytic manifold: (1) the "standard" direct limit topology; (2) the topology induced by the direct limit topology on $\mathscr{E}(K)$; and (3) the topology defined by the seminorms $p^{\omega}_{K,\boldsymbol{a}}$, $K \subseteq \mathsf{M}$ compact, $\boldsymbol{a} \in c_0(\mathbb{Z}_{\geq 0}; \mathbb{R}_{>0})$. We showed in Lemma 5.3 and Theorem 5.5 that these three topologies agree. Now we shall use these constructions to easily arrive at (1) a topology on $\Gamma^{\omega}(\mathsf{E})$ induced by the locally convex topologies on the spaces $\mathscr{G}^{\mathrm{hol},\mathbb{R}}_{K,\overline{\mathsf{E}}}$, $K \subseteq \mathsf{M}$ compact, and (2) seminorms for the topology of $\Gamma^{\omega}(\mathsf{E})$.

For a compact set $K \subseteq \mathsf{M}$ we have an inclusion $i_K : \Gamma^{\omega}(\mathsf{E}) \to \mathscr{G}^{\mathrm{hol},\mathbb{R}}_{K,\overline{\mathsf{E}}}$ defined as follows. If $\xi \in \Gamma^{\omega}(\mathsf{E})$, then ξ admits an holomorphic extension $\overline{\xi}$ defined on a neighbourhood $\overline{\mathcal{U}} \subseteq \overline{\mathsf{M}}$ of M [6, Lemma 5.40]. Since $\overline{\mathcal{U}} \in \mathscr{N}_K$ we define $i_K(\xi) = [\overline{\xi}]_K$. Now we have a compact exhaustion $(K_j)_{j\in\mathbb{Z}_{>0}}$ of M. Since $\mathscr{N}_{K_{j+1}} \subseteq \mathscr{N}_{K_j}$ we have a projection

$$\begin{aligned} \pi_j : \mathscr{G}^{\mathrm{hol},\mathbb{R}}_{K_{j+1},\overline{\mathsf{E}}} &\to \mathscr{G}^{\mathrm{hol},\mathbb{R}}_{K_j,\overline{\mathsf{E}}} \\ [\overline{\xi}]_{K_{j+1}} &\mapsto [\overline{\xi}]_{K_j}. \end{aligned}$$

One can check that, as $\mathbb{R}$-vector spaces, the inverse limit of the inverse family $(\mathscr{G}^{\mathrm{hol},\mathbb{R}}_{K_j,\overline{\mathsf{E}}})_{j\in\mathbb{Z}_{>0}}$ is isomorphic to $\mathscr{G}^{\mathrm{hol},\mathbb{R}}_{\mathsf{M},\overline{\mathsf{E}}}$, the isomorphism being given explicitly by the inclusions

$$\begin{aligned} i_j : \mathscr{G}^{\mathrm{hol},\mathbb{R}}_{\mathsf{M},\overline{\mathsf{E}}} &\to \mathscr{G}^{\mathrm{hol},\mathbb{R}}_{K_j,\overline{\mathsf{E}}} \\ [\overline{\xi}]_{\mathsf{M}} &\mapsto [\overline{\xi}]_{K_j}. \end{aligned}$$

Keeping in mind Lemma 5.2, we then have the inverse limit topology on $\Gamma^{\omega}(\mathsf{E})$ induced by the mappings i_j, $j \in \mathbb{Z}_{>0}$. The topology so defined we call the ***inverse C^{ω}-topology*** for $\Gamma^{\omega}(\mathsf{E})$.

It is now a difficult theorem of Martineau [17, Theorem 1.2(a)] that the direct C^{ω}-topology of Sect. 5.1.3 agrees with the inverse C^{ω}-topology. Therefore, we call the resulting topology the ***C^{ω}-topology***. It is clear from Theorem 5.5 and the preceding inverse limit construction that the seminorms $p^{\omega}_{K,\boldsymbol{a}}$, $K \subseteq \mathsf{M}$ compact, $\boldsymbol{a} \in c_0(\mathbb{Z}_{\geq 0}; \mathbb{R}_{>0})$, define the C^{ω}-topology.

5.3 Properties of the C^{ω}-Topology

To say some relevant things about the C^{ω}-topology, let us first consider the direct limit topology for $\mathscr{G}^{\mathrm{hol},\mathbb{R}}_{K,\overline{\mathsf{E}}}$, $K \subseteq \mathsf{M}$ compact, as this is an important building block for the C^{ω}-topology. First, we recall that a ***strict direct limit*** of locally convex spaces consists of a sequence $(\mathsf{V}_j)_{j\in\mathbb{Z}_{>0}}$ of locally convex spaces that are subspaces of some vector space V, and which have the nesting property $\mathsf{V}_j \subseteq \mathsf{V}_{j+1}$, $j \in \mathbb{Z}_{>0}$. In defining the direct limit topology for $\mathscr{G}^{\mathrm{hol},\mathbb{R}}_{K,\overline{\mathsf{E}}}$ we defined it as a strict direct limit of Banach spaces. Moreover, the restriction mappings from $\Gamma^{\mathrm{hol},\mathbb{R}}_{\mathrm{bdd}}(\overline{\mathsf{E}}|\overline{\mathcal{U}}_j)$ to $\Gamma^{\mathrm{hol},\mathbb{R}}_{\mathrm{bdd}}(\overline{\mathsf{E}}|\overline{\mathcal{U}}_{j+1})$ can be shown to be compact [15, Theorem 8.4]. Direct limits such as these are known as "Silva spaces" or "DFS spaces". Silva spaces have some nice properties, and these provide some of the following attributes for the direct limit topology for $\mathscr{G}^{\mathrm{hol},\mathbb{R}}_{K,\overline{\mathsf{E}}}$.

$\mathscr{G}^{\mathrm{hol},\mathbb{R}}$-1. It is Hausdorff: [19, Theorem 12.1.3].

$\mathscr{G}^{\mathrm{hol},\mathbb{R}}$-2. It is complete: [19, Theorem 12.1.10].

$\mathscr{G}^{\mathrm{hol},\mathbb{R}}$-3. It is not metrisable: [19, Theorem 12.1.8].

$\mathscr{G}^{\mathrm{hol},\mathbb{R}}$-4. It is regular: [15, Theorem 8.4]. This means that every von Neumann bounded subset of $\mathscr{G}^{\mathrm{hol},\mathbb{R}}_{K,\overline{\mathsf{E}}}$ is contained and von Neumann bounded in $\Gamma^{\mathrm{hol},\mathbb{R}}(\overline{\mathsf{E}}|\overline{\mathcal{U}}_j)$ for some $j \in \mathbb{Z}_{>0}$.

$\mathscr{G}^{\mathrm{hol},\mathbb{R}}$-5. It is reflexive: [15, Theorem 8.4].

$\mathscr{G}^{\mathrm{hol},\mathbb{R}}$-6. Its strong dual is a nuclear Fréchet space: [15, Theorem 8.4]. Combined with reflexivity, this means that $\mathscr{G}^{\mathrm{hol},\mathbb{R}}_{K,\overline{\mathsf{E}}}$ is the strong dual of a nuclear Fréchet space.

$\mathscr{G}^{\mathrm{hol},\mathbb{R}}$-7. It is nuclear: [22, Corollary III.7.4].

$\mathscr{G}^{\mathrm{hol},\mathbb{R}}$-8. It is Suslin: This follows from [9, Théorème I.5.1(b)] since $\mathscr{G}^{\mathrm{hol},\mathbb{R}}_{K,\overline{\mathsf{E}}}$ is a strict direct limit of separable Fréchet spaces.

These attributes for the spaces $\mathscr{G}^{\mathrm{hol},\mathbb{R}}_{K,\overline{\mathsf{E}}}$ lead, more or less, to the following attributes of $\Gamma^\omega(\mathsf{E})$.

C^ω-1. It is Hausdorff: It is a union of Hausdorff topologies.

C^ω-2. It is complete: [13, Corollary to Proposition 2.11.3].

C^ω-3. It is not metrisable: It is a union of non-metrisable topologies.

C^ω-4. It is separable: [8, Theorem 16].

C^ω-5. It is nuclear: [22, Corollary III.7.4].

C^ω-6. It is Suslin: Here we note that a countable direct product of Suslin spaces is Suslin [4, Lemma 6.6.5(iii)]. Next we note that the inverse limit is a closed subspace of the direct product [20, Proposition V.19]. Next, closed subspaces of Suslin spaces are Suslin spaces [4, Lemma 6.6.5(ii)]. Therefore, since $\Gamma^\omega(\mathsf{E})$ is the inverse limit of the Suslin spaces $\mathscr{G}^{\mathrm{hol},\mathbb{R}}_{K_j,\overline{\mathsf{E}}}$, $j \in \mathbb{Z}_{>0}$, we conclude that $\Gamma^\omega(\mathsf{E})$ is Suslin.

As we have seen with the CO^∞- and $\mathrm{CO}^{\mathrm{hol}}$-topologies for $\Gamma^\infty(\mathsf{E})$ and $\Gamma^{\mathrm{hol}}(\mathsf{E})$, nuclearity of the C^ω-topology implies that compact subsets of $\Gamma^\omega(\mathsf{E})$ are exactly those that are closed and von Neumann bounded. For von Neumann boundedness, we have the following characterisation.

Lemma 5.6 (Bounded Subsets in the C^ω-Topology). *A subset $\mathcal{B} \subseteq \Gamma^\omega(\mathsf{E})$ is bounded in the von Neumann bornology if and only if the following property holds: for any compact set $K \subseteq \mathsf{M}$ and any $\boldsymbol{a} \in c_0(\mathbb{Z}_{\geq 0}; \mathbb{R}_{>0})$, there exists $C \in \mathbb{R}_{>0}$ such that $p^\omega_{K,\boldsymbol{a}}(\xi) \leq C$ for every $\xi \in \mathcal{B}$.*

5.4 The Weak-$\mathscr{L}$ Topology for Real Analytic Vector Fields

As in the finitely differentiable, Lipschitz, smooth, and holomorphic cases, the above constructions for general vector bundles can be applied to the tangent bundle and the trivial vector bundle $\mathsf{M} \times \mathbb{R}$ to give the ***C^{ω}-topology*** on the space $\Gamma^{\omega}(\mathsf{TM})$ of real analytic vector fields and the space $C^{\omega}(\mathsf{M})$ of real analytic functions. As we have already done in these other cases, we wish to provide a weak characterisation of the C^{ω}-topology for $\Gamma^{\omega}(\mathsf{TM})$. First of all, if $X \in \Gamma^{\omega}(\mathsf{TM})$, then $f \mapsto \mathscr{L}_X f$ is a derivation of $C^{\omega}(\mathsf{M})$. As we have seen, in the holomorphic case this does not generally establish a correspondence between vector fields and derivations, but it does for Stein manifolds. In the real analytic case, Grabowski [10] shows that the map $X \mapsto \mathscr{L}_X$ is indeed an isomorphism of the $\mathbb{R}$-vector spaces of real analytic vector fields and derivations of real analytic functions. Thus the pursuit of a weak description of the C^{ω}-topology for vector fields does not seem to be out of line.

The definition of the weak-$\mathscr{L}$ topology proceeds much as in the smooth and holomorphic cases.

Definition 5.7 (Weak-$\mathscr{L}$ Topology for Space of Real Analytic Vector Fields). For a real analytic manifold M, the ***weak-$\mathscr{L}$ topology*** for $\Gamma^{\omega}(\mathsf{TM})$ is the weakest topology for which the map $X \mapsto \mathscr{L}_X f$ is continuous for every $f \in C^{\omega}(\mathsf{M})$, if $C^{\omega}(\mathsf{M})$ has the C^{ω}-topology. ∘

We now have the following result.

Theorem 5.8 (Weak-$\mathscr{L}$ Characterisation of C^{ω}-Topology for Real Analytic Vector Fields). *For a real analytic manifold* M*, the following topologies for* $\Gamma^{\omega}(\mathsf{TM})$ *agree:*

(i) the C^{ω}-topology;
(ii) the weak-$\mathscr{L}$ topology.

Proof: (i)⊆(ii) As we argued in the corresponding part of the proof of Theorem 3.5, it suffices to show that, for $K \subseteq \mathsf{M}$ compact and for $\boldsymbol{a} \in c_0(\mathbb{Z}_{\geq 0}; \mathbb{R}_{>0})$, there exist compact sets $K_1, \dots, K_r \subseteq \mathsf{M}$, $\boldsymbol{a}_1, \dots, \boldsymbol{a}_r \in c_0(\mathbb{Z}_{\geq 0}; \mathbb{R}_{>0})$, $f^1, \dots, f^r \in C^{\omega}(\mathsf{M})$, and $C_1, \dots, C_r \in \mathbb{R}_{>0}$ such that

$$p^{\omega}_{K,\boldsymbol{a}}(X) \leq C_1 p^{\omega}_{K_1,\boldsymbol{a}_1}(\mathscr{L}_X f^1) + \cdots + C_r p^{\omega}_{K_r,\boldsymbol{a}_r}(\mathscr{L}_X f^r), \qquad X \in \Gamma^{\omega}(\mathsf{TM}).$$

We begin with a simple technical lemma.

Lemma 1 *For each $x \in \mathsf{M}$ there exist $f^1, \dots, f^n \in C^{\omega}(\mathsf{M})$ such that $(\mathrm{d}f^1(x), \dots, \mathrm{d}f^n(x))$ is a basis for $\mathsf{T}^*_x\mathsf{M}$.*

Proof: We are supposing, of course, that the connected component of M containing x has dimension n. There are many ways to prove this lemma, including applying Cartan's Theorem A to the sheaf of real analytic functions on M. We shall prove the lemma by embedding M in $\mathbb{R}^N$ by the embedding theorem of Grauert [11]. Thus we have a proper real analytic embedding $\iota_{\mathsf{M}} \colon \mathsf{M} \to \mathbb{R}^N$. Let $g^1, \dots, g^N \in C^{\omega}(\mathbb{R}^N)$ be the coordinate functions. Then we have a surjective linear map

$$\begin{aligned}\sigma_x\colon \mathbb{R}^N &\to \mathsf{T}_x^*\mathsf{M}\\ (c_1,\dots,c_N) &\mapsto \sum_{j=1}^N c_j \mathrm{d}(\iota_{\mathsf{M}}^* g^j)(x).\end{aligned}$$

Let $\boldsymbol{c}^1,\dots,\boldsymbol{c}^n \in \mathbb{R}^N$ be a basis for a complement of $\ker(\sigma_x)$. Then the functions

$$f^j = \sum_{k=1}^N c_k^j \iota_{\mathsf{M}}^* g^k$$

have the desired property. ▽

We assume that M has a well-defined dimension n. This assumption can easily be relaxed. We use the notation

$$p'^{\omega}_{K,\boldsymbol{a}}(f) = \sup\Big\{\frac{a_0 a_1 \cdots a_{|I|}}{I!}|\boldsymbol{D}^I f(\boldsymbol{x})| \;\Big|\; \boldsymbol{x}\in K,\ I \in \mathbb{Z}_{\geq 0}^n\Big\}$$

for a function $f \in \mathrm{C}^\omega(\mathcal{U})$ defined on an open subset of $\mathbb{R}^n$ and with $K \subseteq \mathcal{U}$ compact. We shall also use this local coordinate notation for seminorms of local representatives of vector fields. Let $K \subseteq \mathsf{M}$ be compact and let $\boldsymbol{a} \in \mathrm{c}_0(\mathbb{Z}_{\geq 0};\mathbb{R}_{>0})$. Let $x \in K$ and let $(\mathcal{U}_x,\phi_x)$ be a chart for M about x with the property that the coordinate functions x^j, $j \in \{1,\dots,n\}$, are restrictions to $\mathcal{U}_x$ of globally defined real analytic functions f_x^j, $j \in \{1,\dots,n\}$, on M. This is possible by the lemma above. Let $\boldsymbol{X}\colon \phi_x(\mathcal{U}_x) \to \mathbb{R}^n$ be the local representative of $X \in \Gamma^\omega(\mathsf{M})$. Then, in a neighbourhood of the closure of a relatively compact neighbourhood $\mathcal{V}_x \subseteq \mathcal{U}_x$ of x, we have $\mathscr{L}_X f_x^j = X^j$, the jth component of X. By Lemma 2.5, there exist $C_x, \sigma_x \in \mathbb{R}_{>0}$ such that

$$\|j_m X(y)\|_{\overline{\mathbb{G}}_m} \leq C_x \sigma_x^{-m} \sup\Big\{\frac{1}{I!}|\boldsymbol{D}^I X^j(\phi_x(y))| \;\Big|\; |I| \leq m,\ j \in \{1,\dots,n\}\Big\}$$

for $m \in \mathbb{Z}_{\geq 0}$ and $y \in \mathrm{cl}(\mathcal{V}_x)$. By equivalence of the ℓ^1 and ℓ^∞-norms for $\mathbb{R}^n$, there exists $C \in \mathbb{R}_{>0}$ such that

$$\begin{aligned}&\sup\Big\{\frac{1}{I!}|\boldsymbol{D}^I X^j(\phi_x(y))| \;\Big|\; |I| \leq m,\ j \in \{1,\dots,n\}\Big\}\\ &\qquad\leq C \sum_{j=1}^n \sup\Big\{\frac{1}{I!}|\boldsymbol{D}^I(\mathscr{L}_X f_x^j)(\phi_x(y))| \;\Big|\; |I| \leq m\Big\}\end{aligned}$$

for $m \in \mathbb{Z}_{\geq 0}$ and $y \in \mathrm{cl}(\mathcal{V}_x)$. Another application of Lemma 2.5 gives $B_x, r_x \in \mathbb{R}_{>0}$ such that

$$\sup\Big\{\frac{1}{I!}|\boldsymbol{D}^I(\mathscr{L}_X f_x^j)(\phi_x(y))| \;\Big|\; |I| \leq m,\ j \in \{1,\dots,n\}\Big\} \leq B_x r_x^{-m} \|j_m(\mathscr{L}_X f_x^j)(y)\|$$

for $m \in \mathbb{Z}_{\geq 0}$, $j \in \{1, \dots, n\}$, and $y \in \mathrm{cl}(\mathcal{V}_x)$. Combining the preceding three estimates and renaming constants gives

$$\|j_m X(y)\|_{\overline{\mathbb{G}}_m} \leq \sum_{j=1}^{n} C_x \sigma_x^{-m} \|j_m(\mathscr{L}_X f_x^j(\phi_x(y)))\|_{\overline{\mathbb{G}}_m}$$

for $m \in \mathbb{Z}_{\geq 0}$ and $y \in \mathrm{cl}(\mathcal{V}_x)$. Define

$$\boldsymbol{b}_x = (b_m)_{m \in \mathbb{Z}_{\geq 0}} \in \mathrm{c}_0(\mathbb{Z}_{\geq 0}; \mathbb{R}_{>0})$$

by $b_0 = C_x a_0$ and $b_m = \sigma_x^{-1} a_m$, $m \in \mathbb{Z}_{>0}$. Therefore,

$$a_0 a_1 \cdots a_m \|j_m X(y)\|_{\overline{\mathbb{G}}_m} \leq \sum_{j=1}^{n} b_0 b_1 \cdots b_m \|j_m(\mathscr{L}_X f_x^j(\phi_x(y)))\|_{\overline{\mathbb{G}}_m}$$

for $m \in \mathbb{Z}_{\geq 0}$ and $y \in \mathrm{cl}(\mathcal{V}_x)$. Taking supremums over $y \in \mathrm{cl}(\mathcal{V}_x)$ and $m \in \mathbb{Z}_{\geq 0}$ on the right gives

$$a_0 a_1 \cdots a_m \|j_m X(y)\|_{\overline{\mathbb{G}}_m} \leq \sum_{j=1}^{n} p^{\omega}_{\mathrm{cl}(\mathcal{V}_x), \boldsymbol{b}_x}(\mathscr{L}_X f_x^j), \qquad m \in \mathbb{Z}_{\geq 0},\ y \in \mathrm{cl}(\mathcal{V}_x).$$

Let $x_1, \dots, x_k \in K$ be such that $K \subseteq \cup_{j=1}^{k} \mathcal{V}_{x_j}$, let $f^1, \dots, f^{kn}$ be the list of functions

$$f_{x_1}^1, \dots, f_{x_1}^n, \dots, f_{x_k}^1, \dots, f_{x_k}^n,$$

and let $\boldsymbol{b}_1, \dots, \boldsymbol{b}_{kn} \in \mathrm{c}_0(\mathbb{Z}_{\geq 0}; \mathbb{R}_{>0})$ be the list of sequences

$$\underbrace{\boldsymbol{b}_{x_1}, \dots, \boldsymbol{b}_{x_1}}_{n \text{ times}}, \dots, \underbrace{\boldsymbol{b}_{x_k}, \dots, \boldsymbol{b}_{x_k}}_{n \text{ times}}.$$

If $x \in K$, then $x \in \mathcal{V}_{x_j}$ for some $j \in \{1, \dots, k\}$ and so

$$a_0 a_1 \cdots a_m \|j_m X(x)\|_{\overline{\mathbb{G}}_m} \leq \sum_{j=1}^{kn} p^{\omega}_{K, \boldsymbol{b}_j}(\mathscr{L}_X f^j),$$

and this part of the lemma follows upon taking the supremum over $x \in K$ and $m \in \mathbb{Z}_{\geq 0}$.

(ii)⊆(i) Here, as in the proof of the corresponding part of Theorem 3.5, it suffices to show that, for every $f \in \mathrm{C}^{\omega}(\mathsf{M})$, the map $\mathscr{L}_f \colon X \mapsto \mathscr{L}_X f$ is continuous from $\Gamma^{\omega}(\mathsf{TM})$ with the C^{ω}-topology to $\mathrm{C}^{\omega}(\mathsf{M})$ with the C^{ω}-topology.

We shall use the direct C^{ω}-topology to show this. Thus we work with an holomorphic manifold $\overline{\mathsf{M}}$ that is a complexification of M, as described in Sect. 5.1.1. We recall that $\mathscr{N}_{\mathsf{M}}$ denotes the directed set of neighbourhoods of M in $\overline{\mathsf{M}}$, and that the set

$\mathscr{S}_M$ of Stein neighbourhoods is cofinal in $\mathscr{N}_M$. As we saw in Sect. 5.1.3, for $\overline{\mathscr{U}} \in \mathscr{N}_M$, we have mappings

$$r_{\overline{\mathscr{U}},M}: \Gamma^{hol,\mathbb{R}}(T\overline{\mathscr{U}}) \to \Gamma^\omega(TM)$$

$$\overline{X} \mapsto \overline{X}|M$$

and

$$r_{\overline{\mathscr{U}},M}: C^{hol,\mathbb{R}}(\overline{\mathscr{U}}) \to C^\omega(M)$$

$$\overline{f} \mapsto \overline{f}|M,$$

making an abuse of notation by using $r_{\overline{\mathscr{U}},M}$ for two different things, noting that context will make it clear which we mean. For $K \subseteq M$ compact, we also have the mapping

$$i_{M,K}: C^\omega(M) \to \mathscr{C}^{hol,\mathbb{R}}_{K,\overline{M}}$$

$$f \mapsto [\overline{f}]_K,$$

where $\mathscr{C}^{hol,\mathbb{R}}_{K,\overline{M}}$ is the set of germs of holomorphic functions about K, i.e., the usual construction of Sect. 5.2.1, but in the special case of $E = M \times \mathbb{R}$. The C^ω-topology is the final topology induced by the mappings $r_{\overline{\mathscr{U}},M}$. As such, by [13, Proposition 2.12.1], the map $\mathscr{L}_f$ is continuous if and only if $\mathscr{L}_f \circ r_{\overline{\mathscr{U}},M}$ for every $\overline{\mathscr{U}} \in \mathscr{N}_M$. Thus let $\overline{\mathscr{U}} \in \mathscr{N}_M$. To show that $\mathscr{L}_f \circ r_{\overline{\mathscr{U}},M}$ is continuous, it suffices by [13, Sect. 2.11] to show that $i_{M,K} \circ \mathscr{L}_f \circ r_{\overline{\mathscr{U}},M}$ is continuous for every compact $K \subseteq M$. Next, there is $\overline{\mathscr{U}} \supseteq \overline{\mathscr{S}} \in \mathscr{S}_M$ so that f admits an holomorphic extension $\overline{f}$ to $\overline{\mathscr{S}}$. The following diagram shows how this all fits together.

$$\Gamma^{hol,\mathbb{R}}(T\overline{\mathscr{U}}) \xrightarrow{r_{\overline{\mathscr{U}},\overline{\mathscr{S}}}} \Gamma^{hol,\mathbb{R}}(T\overline{\mathscr{S}}) \xrightarrow{r_{\overline{\mathscr{S}},M}} \Gamma^\omega(TM)$$

$$C^{hol,\mathbb{R}}(\overline{\mathscr{S}}) \xrightarrow{r_{\overline{\mathscr{S}},M}} C^\omega(M) \xrightarrow{i_{M,K}} \mathscr{C}^{hol,\mathbb{R}}_{K,\overline{M}}$$

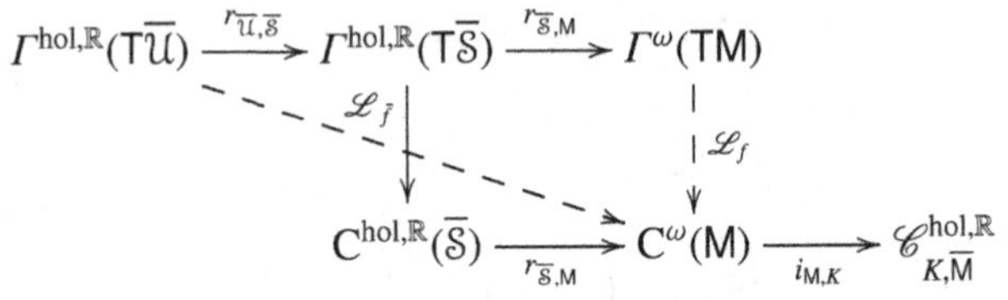

The dashed arrows signify maps whose continuity is a priori unknown to us. The diagonal dashed arrow is the one whose continuity we must verify to ascertain the continuity of the vertical dashed arrow. It is a simple matter of checking definitions to see that the diagram commutes. By Theorem 4.5, we have that $\mathscr{L}_{\overline{f}}: \Gamma^{hol,\mathbb{R}}(T\overline{\mathscr{S}}) \to C^{hol,\mathbb{R}}(\overline{\mathscr{S}})$ is continuous (keeping Remark 5.1 in mind). We deduce that, since

$$i_{M,K} \circ \mathscr{L}_f \circ r_{\overline{\mathscr{U}},M} = i_{M,K} \circ r_{\overline{\mathscr{S}},M} \circ \mathscr{L}_{\overline{f}} \circ r_{\overline{\mathscr{U}},\overline{\mathscr{S}}},$$

$i_{M,K} \circ \mathscr{L}_f \circ r_{\overline{\mathscr{U}},M}$ is continuous for every $\overline{\mathscr{U}} \in \mathscr{N}_M$ and for every compact $K \subseteq M$, as desired.   □

As in the smooth and holomorphic cases, we can prove the equivalence of various topological notions between the weak-$\mathscr{L}$ and usual topologies.

Corollary 5.9 (Weak-$\mathscr{L}$ Characterisations of Boundedness, Continuity, Measurability, and Integrability for the C^ω-Topology). *Let* M *be a real analytic manifold, let* $(\mathcal{X}, \mathscr{O})$ *be a topological space, let* $(\mathcal{T}, \mathscr{M})$ *be a measurable space, and let* $\mu \colon \mathscr{M} \to \overline{\mathbb{R}}_{\geq 0}$ *be a finite measure. The following statements hold:*

(i) *a subset* $\mathcal{B} \subseteq \Gamma^\omega(\mathsf{TM})$ *is bounded in the von Neumann bornology if and only if it is weak-*$\mathscr{L}$ *bounded in the von Neumann bornology;*
(ii) *a map* $\Phi \colon \mathcal{X} \to \Gamma^\omega(\mathsf{TM})$ *is continuous if and only if it is weak-*$\mathscr{L}$ *continuous;*
(iii) *a map* $\Psi \colon \mathcal{T} \to \Gamma^\omega(\mathsf{TM})$ *is measurable if and only if it is weak-*$\mathscr{L}$ *measurable;*
(iv) *a map* $\Psi \colon \mathcal{T} \to \Gamma^\omega(\mathsf{TM})$ *is Bochner integrable if and only if it is weak-*$\mathscr{L}$ *Bochner integrable.*

Proof: The fact that $\{\mathscr{L}_f \mid f \in C^\omega(\mathsf{M})\}$ contains a countable point separating subset follows from combining the lemma from the proof of Theorem 5.8 with the proof of the corresponding assertion in Corollary 3.6. Since $\Gamma^\omega(\mathsf{TM})$ is complete, separable, and Suslin, and since $C^\omega(\mathsf{M})$ is Suslin by properties C^ω-2, C^ω-4, and C^ω-6 above, the corollary follows from Lemma 3.3, taking "$\mathsf{U} = \Gamma^\omega(\mathsf{TM})$", "$\mathsf{V} = C^\omega(\mathsf{M})$", and "$\mathscr{A} = \{\mathscr{L}_f \mid f \in C^\omega(\mathsf{M})\}$". □

References

1. Agrachev, A.A., Gamkrelidze, R.V.: The exponential representation of flows and the chronological calculus. Mathematics of the USSR-Sbornik **107**(4), 467–532 (1978)
2. Baernstein II, A.: Representation of holomorphic functions by boundary integrals. Transactions of the American Mathematical Society **160**, 27–37 (1971)
3. Bierstedt, K.D.: An introduction to locally convex inductive limits. In: Functional Analysis and its Applications, ICPAM Lecture Notes, pp. 35–133. World Scientific, Singapore/New Jersey/London/Hong Kong (1988)
4. Bogachev, V.I.: Measure Theory, vol. 2. Springer-Verlag, New York/Heidelberg/Berlin (2007)
5. Borel, E.: Sur quelles points de la théorie des fonctions. Annales Scientifiques de l'École Normale Supérieure. Quatrième Série **12**(3), 44 (1895)
6. Cieliebak, K., Eliashberg, Y.: From Stein to Weinstein and Back: Symplectic Geometry of Affine Complex Manifolds. No. 59 in American Mathematical Society Colloquium Publications. American Mathematical Society, Providence, RI (2012)
7. Conway, J.B.: A Course in Functional Analysis, 2 edn. No. 96 in Graduate Texts in Mathematics. Springer-Verlag, New York/Heidelberg/Berlin (1985)
8. Domański, P.: Notes on real analytic functions and classical operators. In: O. Blasco, J. Bonet, J. Calabuig, D. Jornet (eds.) Proceedings of the Third Winter School in Complex Analysis and Operator Theory, *Contemporary Mathematics*, vol. 561, pp. 3–47. American Mathematical Society, Providence, RI (2010)
9. Fernique, X.: Processus linéares, processus général. Université de Grenoble. Annales de l'Institut Fourier **17**(1), 1–92 (1967)
10. Grabowski, J.: Derivations of Lie algebras of analytic vector fields. Compositio Mathematica **43**(2), 239–252 (1981)
11. Grauert, H.: On Levi's problem and the imbedding of real-analytic manifolds. Annals of Mathematics. Second Series **68**, 460–472 (1958)
12. Groethendieck, A.: Topological Vector Spaces. Notes on Mathematics and its Applications. Gordon & Breach Science Publishers, New York (1973)
13. Horváth, J.: Topological Vector Spaces and Distributions. Vol. I. Addison Wesley, Reading, MA (1966)

14. Jarchow, H.: Locally Convex Spaces. Mathematical Textbooks. Teubner, Leipzig (1981)
15. Kriegl, A., Michor, P.W.: The Convenient Setting of Global Analysis. No. 57 in American Mathematical Society Mathematical Surveys and Monographs. American Mathematical Society, Providence, RI (1997)
16. Lang, S.: Algebra. No. 211 in Graduate Texts in Mathematics. Springer-Verlag, New York/-Heidelberg/Berlin (2002)
17. Martineau, A.: Sur la topologie des espaces de fonctions holomorphes. Mathematische Annalen **163**, 62–88 (1966)
18. Mujica, J.: A Banach–Dieudonné theorem for germs of holomorphic functions. Journal of Functional Analysis **57**(1), 31–48 (1984)
19. Narici, L., Beckenstein, E.: Topological Vector Spaces, 2 edn. Pure and Applied Mathematics. CRC Press, Boca Raton, FL (2010)
20. Robertson, A.P., Robertson, W.: Topological Vector Spaces, 2 edn. No. 53 in Cambridge Tracts in Mathematics. Cambridge University Press, New York/Port Chester/Melbourne/Sydney (1980)
21. Saunders, D.J.: The Geometry of Jet Bundles. No. 142 in London Mathematical Society Lecture Note Series. Cambridge University Press, New York/Port Chester/Melbourne/Sydney (1989)
22. Schaefer, H.H., Wolff, M.P.: Topological Vector Spaces, 2 edn. No. 3 in Graduate Texts in Mathematics. Springer-Verlag, New York/Heidelberg/Berlin (1999)
23. Stein, E.M., Weiss, G.: Introduction to Fourier Analysis on Euclidean Space. No. 32 in Princeton Mathematical Series. Princeton University Press, Princeton, NJ (1971)
24. Vogt, D.: A fundamental system of seminorms for $A(K)$ (2013). URL http://arxiv.org/abs/1309.6292v1. ArXiv:1309.6292v1 [math.FA]
25. Whitney, H., Bruhat, F.: Quelques propriétés fondamentales des ensembles analytiques-réels. Commentarii Mathematici Helvetici **33**, 132–160 (1959)

Chapter 6
Time-Varying Vector Fields

In this chapter we consider time-varying vector fields. The ideas in this chapter originate (for us) with the paper of Agrachev and Gamkrelidze [2], and are nicely summarised in the more recent book by Agrachev and Sachkov [3], at least in the smooth case. A geometric presentation of some of the constructions can be found in the paper of Sussmann [17], again in the smooth case, and Sussmann also considers regularity less than smooth, e.g., finitely differentiable or Lipschitz. There is some consideration of the real analytic case in [2], but this consideration is restricted to real analytic vector fields admitting a bounded holomorphic extension to a fixed-width neighbourhood of $\mathbb{R}^n$ in $\mathbb{C}^n$. One of our results, the rather nontrivial Theorem 6.25, is that this framework of [2] is sufficient for the purposes of local analysis. However, our treatment of the real analytic case is global, general, and comprehensive. To provide some context for our novel treatment of the real analytic case, we treat the smooth case in some detail, even though the results are probably mostly known. (However, we should say that, even in the smooth case, we could not find precise statements with proofs of some of the results we give.) We also treat the finitely differentiable and Lipschitz cases, so our theory also covers the "standard" Carathéodory existence and uniqueness theorem for time-varying ordinary differential equations, e.g., [16, Theorem 54]. We also consider holomorphic time-varying vector fields, as these have a relationship to real analytic time-varying vector fields that is sometimes useful to exploit.

One of the unique facets of our presentation is that we fully explain the rôle of the topologies developed in Chaps. 3, 4, and 5. Indeed, one way to understand the principal results of this chapter is that they show that the usual pointwise—in state and time—conditions placed on vector fields to regulate the character of their flows can be profitably phrased in terms of topologies for spaces of vector fields. While this idea is not entirely new—it is implicit in the approach of [2]—we do develop it comprehensively and in new directions.

While our principal interest is in vector fields, and also in functions, it is convenient to conduct much of the development for general vector bundles, subsequently specialising to vector fields and functions.

S. Jafarpour, A.D. Lewis, *Time-Varying Vector Fields and Their Flows*,
SpringerBriefs in Mathematics, DOI 10.1007/978-3-319-10139-2_6

6.1 The Smooth Case

Throughout this section we will work with a smooth vector bundle $\pi\colon \mathsf{E} \to \mathsf{M}$ with a linear connection ∇^0 on E, an affine connection ∇ on M, a fibre metric $\mathbb{G}_0$ on E, and a Riemannian metric $\mathbb{G}$ on M. This defines the fibre norms $\|\cdot\|_{\overline{\mathbb{G}}_m}$ on $\mathsf{J}^m\mathsf{E}$ and seminorms $p^{\infty}_{K,m}$, $K \subseteq \mathsf{M}$ compact, $m \in \mathbb{Z}_{\geq 0}$, on $\Gamma^{\infty}(\mathsf{E})$ as in Sect. 3.1.

Definition 6.1 (Smooth Carathéodory Section). Let $\pi\colon \mathsf{E} \to \mathsf{M}$ be a smooth vector bundle and let $\mathbb{T} \subseteq \mathbb{R}$ be an interval. A ***Carathéodory section of class*** C^{∞} of E is a map $\xi\colon \mathbb{T} \times \mathsf{M} \to \mathsf{E}$ with the following properties:
(i) $\xi(t, x) \in \mathsf{E}_x$ for each $(t, x) \in \mathbb{T} \times \mathsf{M}$;
(ii) for each $t \in \mathbb{T}$, the map $\xi_t\colon \mathsf{M} \to \mathsf{E}$ defined by $\xi_t(x) = \xi(t, x)$ is of class C^{∞};
(iii) for each $x \in \mathsf{M}$, the map $\xi^x\colon \mathbb{T} \to \mathsf{E}$ defined by $\xi^x(t) = \xi(t, x)$ is Lebesgue measurable.

We shall call $\mathbb{T}$ the ***time-domain*** for the section. By $\mathrm{CF}\Gamma^{\infty}(\mathbb{T}; \mathsf{E})$ we denote the set of Carathéodory sections of class C^{∞} of E. ○

Note that the curve $t \mapsto \xi(t, x)$ is in the finite-dimensional vector space E_x, and so Lebesgue measurability of this is unambiguously defined, e.g., by choosing a basis and asking for Lebesgue measurability of the components with respect to this basis.

Now we put some conditions on the time dependence of the derivatives of the section.

Definition 6.2 (Locally Integrally C^{∞}-Bounded and Locally Essentially C^{∞}-Bounded Sections). Let $\pi\colon \mathsf{E} \to \mathsf{M}$ be a smooth vector bundle and let $\mathbb{T} \subseteq \mathbb{R}$ be an interval. A Carathéodory section $\xi\colon \mathbb{T} \times \mathsf{M} \to \mathsf{E}$ of class C^{∞} is
(i) ***locally integrally*** C^{∞}***-bounded*** if, for every compact set $K \subseteq \mathsf{M}$ and every $m \in \mathbb{Z}_{\geq 0}$, there exists $g \in \mathrm{L}^1_{\mathrm{loc}}(\mathbb{T}; \mathbb{R}_{\geq 0})$ such that

$$\|j_m\xi_t(x)\|_{\overline{\mathbb{G}}_m} \leq g(t), \qquad (t, x) \in \mathbb{T} \times K,$$

and is
(ii) ***locally essentially*** C^{∞}***-bounded*** if, for every compact set $K \subseteq \mathsf{M}$ and every $m \in \mathbb{Z}_{\geq 0}$, there exists $g \in \mathrm{L}^{\infty}_{\mathrm{loc}}(\mathbb{T}; \mathbb{R}_{\geq 0})$ such that

$$\|j_m\xi_t(x)\|_{\overline{\mathbb{G}}_m} \leq g(t), \qquad (t, x) \in \mathbb{T} \times K.$$

The set of locally integrally C^{∞}-bounded sections of E with time-domain $\mathbb{T}$ is denoted by $\mathrm{LI}\Gamma^{\infty}(\mathbb{T}, \mathsf{E})$ and the set of locally essentially C^{∞}-bounded sections of E with time-domain $\mathbb{T}$ is denoted by $\mathrm{LB}\Gamma^{\infty}(\mathbb{T}; \mathsf{E})$. ○

Note that $\mathrm{LB}\Gamma^{\infty}(\mathbb{T}; \mathsf{M}) \subseteq \mathrm{LI}\Gamma^{\infty}(\mathbb{T}; \mathsf{M})$, precisely because locally essentially bounded functions (in the usual sense) are locally integrable (in the usual sense).

We note that our definitions differ from those in [2, 3, 17]. The form of the difference is our use of connections and jet bundles, aided by Lemma 2.1. In [2] the presentation is developed on Euclidean spaces, and so the geometric treatment we give here is not necessary. (One way of understanding why it is not necessary is that Euclidean space has a canonical flat connection in which the decomposition of Lemma 2.1 becomes the usual decomposition of derivatives by their order.) In [3] the treatment is on manifolds, and the seminorms are defined by an embedding of the manifold in Euclidean space by Whitney's Embedding Theorem [20]. Also, Agrachev and Sachkov [3] use the weak-$\mathscr{L}$ topology in the case of vector fields, but we have seen that this is the same as the usual topology (Theorem 3.5). In [17] the characterisation of Carathéodory functions uses Lie differentiation by smooth vector fields, and the locally convex topology for $\Gamma^\infty(\mathsf{TM})$ is not explicitly considered, although it is implicit in Sussmann's constructions. Sussmann also takes a weak-$\mathscr{L}$ approach to characterising properties of time-varying vector fields. In any case, all approaches can be tediously shown to be equivalent once the relationships are understood. An advantage of the approach we use here is that it does not require coordinate charts or embeddings to write the seminorms, and it makes the seminorms explicit, rather than implicitly present. The disadvantage of our approach is the added machinery and complication of connections and our jet bundle decomposition.

The following characterisation of Carathéodory sections and their relatives is useful and insightful.

Theorem 6.3 (Topological Characterisation of Smooth Carathéodory Sections). *Let $\pi\colon \mathsf{E} \to \mathsf{M}$ be a smooth vector bundle and let $\mathbb{T} \subseteq \mathbb{R}$ be an interval. For a map $\xi\colon \mathbb{T} \times \mathsf{M} \to \mathsf{E}$ satisfying $\xi(t, x) \in \mathsf{E}_x$ for each $(t, x) \in \mathbb{T} \times \mathsf{M}$, the following two statements are equivalent:*

(i) $\xi \in \mathrm{CF}\Gamma^\infty(\mathbb{T}; \mathsf{E})$;

(ii) the map $\mathbb{T} \ni t \mapsto \xi_t \in \Gamma^\infty(\mathsf{E})$ is measurable,

the following two statements are equivalent:

(iii) $\xi \in \mathrm{LI}\Gamma^\infty(\mathbb{T}; \mathsf{E})$;

(iv) the map $\mathbb{T} \ni t \mapsto \xi_t \in \Gamma^\infty(\mathsf{E})$ is measurable and locally Bochner integrable,

and the following two statements are equivalent:

(v) $\xi \in \mathrm{LB}\Gamma^\infty(\mathbb{T}; \mathsf{E})$;

(vi) the map $\mathbb{T} \ni t \mapsto \xi_t \in \Gamma^\infty(\mathsf{E})$ is measurable and locally essentially von Neumann bounded.

Proof: It is illustrative, especially since we will refer to this proof at least three times subsequently, to understand the general framework of the proof. Much of the argument has already been carried out in a more general setting in Lemma 3.3.

So we let V be a locally convex topological vector space over $\mathbb{F} \in \{\mathbb{R}, \mathbb{C}\}$, let $(\mathfrak{T}, \mathscr{M})$ be a measurable space, and let $\Psi\colon \mathfrak{T} \to \mathsf{V}$. Let us first characterise measurability of Ψ. We use here the results of Thomas [18] who studies integrability for functions taking values in locally convex Suslin spaces. Thus we assume that V is a Hausdorff Suslin space (as is the case for all spaces of interest to us in this work). We let V' denote the topological dual of V. A subset $S \subseteq \mathsf{V}'$ is ***point separating*** if, for distinct $v_1, v_2 \in \mathsf{V}$, there exists $\alpha \in \mathsf{V}'$ such that $\alpha(v_1) \neq \alpha(v_2)$. Thomas proves the following result as his Theorem 1, and whose proof we provide, as it is straightforward and shows where the (not so straightforward) properties of Suslin spaces are used.

Lemma 1 *Let* V *be a Hausdorff, Suslin, locally convex topological vector space over* $\mathbb{F} \in \{\mathbb{R}, \mathbb{C}\}$, *let* $(\mathfrak{T}, \mathscr{M})$ *be a measurable space, and let* $\Psi : \mathfrak{T} \to \mathsf{V}$. *If* $S \subseteq \mathsf{V}'$ *is point separating, then* Ψ *is measurable if and only if* $\alpha \circ \Psi$ *is measurable for every* $\alpha \in S$.

Proof: If Ψ is measurable, then it is obvious that $\alpha \circ \Psi$ is measurable for every $\alpha \in \mathsf{V}'$ since such α are continuous.

Conversely, suppose that $\alpha \circ \Psi$ is measurable for every $\alpha \in S$. First of all, locally convex topological vector spaces are completely regular if they are Hausdorff [14, p. 16]. Therefore, by [5, Theorem 6.7.7], there is a countable subset of S that is point separating, so we may as well suppose that S is countable. We are now in the same framework as Lemma 3.3(iii), and the proof there applies by taking "$\mathsf{U} = \mathsf{V}$", "$\mathsf{V} = \mathbb{F}$", and "$\mathscr{A} = S$". ▽

The preceding lemma will allow us to characterise measurability. Let us now consider integrability.

Lemma 2 *Let* V *be a complete separable locally convex topological vector space over* $\mathbb{F} \in \{\mathbb{R}, \mathbb{C}\}$ *and let* $(\mathfrak{T}, \mathscr{M}, \mu)$ *be a finite measure space. A measurable function* $\Psi : \mathfrak{T} \to \mathsf{V}$ *is Bochner integrable if and only if* $p \circ \Psi$ *is integrable for every continuous seminorm* p *for* V.

Proof: It follows from [4, Theorems 3.2, 3.3] that Ψ is integrable if $p \circ \Psi$ is integrable for every continuous seminorm p. Conversely, if Ψ is integrable, it is implied that Ψ is Bochner approximable, and so, by [4, Theorem 3.2], we have that $p \circ \Psi$ is integrable for every continuous seminorm p. ▽

(i) $\iff$ (ii) For $x \in \mathsf{M}$ and $\alpha_x \in \mathsf{E}_x^*$, define $\mathrm{ev}_{\alpha_x} : \Gamma^\infty(\mathsf{E}) \to \mathbb{R}$ by $\mathrm{ev}_{\alpha_x}(\xi) = \langle \alpha_x; \xi(x) \rangle$. Clearly ev_{α_x} is $\mathbb{R}$-linear. We claim that ev_{α_x} is continuous. Indeed, for a directed set $(I, \preceq)$ and a net $(\xi)_{i \in I}$ converging to ξ,[1] we have

$$\lim_{i \in I} \mathrm{ev}_{\alpha_x}(\xi_i) = \lim_{i \in I} \alpha_x(\xi_i(x)) = \alpha_x\Big(\lim_{i \in I} \xi_i(x)\Big) = \alpha_x(\xi(x)) = \mathrm{ev}_{\alpha_x}(\xi),$$

using the fact that convergence in the CO^∞-topology implies pointwise convergence. It is obvious that the continuous linear functions ev_{α_x}, $\alpha_x \in \mathsf{E}^*$, are point separating. We now recall from property CO^∞-6 for the smooth CO^∞-topology that $\Gamma^\infty(\mathsf{E})$ is a Suslin space with the CO^∞-topology. Therefore, by the first lemma above, it follows that $t \mapsto \xi_t$ is measurable if and only if $t \mapsto \mathrm{ev}_{\alpha_x}(\xi_t) = \langle \alpha_x; \xi_t(x) \rangle$ is measurable for every $\alpha_x \in \mathsf{E}^*$. On the other hand, this is equivalent to $t \mapsto \xi_t(x)$ being measurable for every $x \in \mathsf{M}$ since $t \mapsto \xi_t(x)$ is a curve in the finite-dimensional vector space E_x. Finally, note that it is implicit in the statement of (ii) that ξ_t is smooth, and this part of the proposition follows easily from these observations.

(iii) $\iff$ (iv) Let $\mathbb{T}' \subseteq \mathbb{T}$ be compact.

[1] Since $\Gamma^\infty(\mathsf{E})$ is metrisable, it suffices to use sequences. However, we shall refer to this argument when we do not use metrisable spaces, so it is convenient to have the general argument here.

First suppose that $\xi \in \mathrm{LI}\Gamma^\infty(\mathbb{T};\mathsf{E})$. By definition of locally integrally C^∞-bounded, for each compact $K \subseteq \mathsf{M}$ and $m \in \mathbb{Z}_{\geq 0}$, there exists $g \in \mathrm{L}^1(\mathbb{T}';\mathbb{R}_{\geq 0})$ such that

$$\|j_m\xi_t(x)\|_{\overline{\mathbb{G}}_m} \leq g(t), \qquad (t,x) \in \mathbb{T}' \times K.$$

Therefore, $p^\infty_{K,m}(\xi_t) \leq g(t)$ for $t \in \mathbb{T}'$. Note that continuity of $p^\infty_{K,m}$ implies that $t \mapsto p^\infty_{K,m}(\xi_t)$ is measurable. Therefore,

$$\int_{\mathbb{T}'} p^\infty_{K,m}(\xi_t)\,\mathrm{d}t < \infty, \qquad K \subseteq \mathsf{M} \text{ compact}, \ m \in \mathbb{Z}_{\geq 0}.$$

Since $\Gamma^\infty(\mathsf{E})$ is complete and separable, it now follows from the second lemma above that $t \mapsto \xi_t$ is Bochner integrable on $\mathbb{T}'$. That is, since $\mathbb{T}'$ is arbitrary, $t \mapsto \xi_t$ is locally Bochner integrable.

Next suppose that $t \mapsto \xi_t$ is Bochner integrable on $\mathbb{T}$. By the second lemma above,

$$\int_{\mathbb{T}'} p^\infty_{K,m}(\xi_t)\,\mathrm{d}t < \infty, \qquad K \subseteq \mathsf{M} \text{ compact}, \ m \in \mathbb{Z}_{\geq 0}.$$

Therefore, since

$$\|j_m\xi_t(x)\|_{\overline{\mathbb{G}}_m} \leq p^\infty_{K,m}(\xi_t), \qquad (t,x) \in \mathbb{T}' \times K,$$

we conclude that ξ is locally integrally C^∞-bounded since $\mathbb{T}'$ is arbitrary.

(v) $\Longleftrightarrow$ (vi) We recall our discussion of von Neumann bounded sets in locally convex topological vector spaces preceding Lemma 3.1 above. With this in mind and using Lemma 3.1, this part of the theorem follows immediately. □

Note that Theorem 6.3 applies, in particular, to vector fields and functions, giving the classes $\mathrm{CF}^\infty(\mathbb{T};\mathsf{M})$, $\mathrm{LIC}^\infty(\mathbb{T};\mathsf{M})$, and $\mathrm{LBC}^\infty(\mathbb{T};\mathsf{M})$ of functions, and the classes $\mathrm{CF}\Gamma^\infty(\mathbb{T};\mathsf{TM})$, $\mathrm{LI}\Gamma^\infty(\mathbb{T};\mathsf{TM})$, and $\mathrm{LB}\Gamma^\infty(\mathbb{T};\mathsf{TM})$ of vector fields. Noting that we have the alternative weak-$\mathscr{L}$ characterisation of the CO^∞-topology, we can summarise the various sorts of measurability, integrability, and boundedness for smooth time-varying vector fields as follows. In the statement of the result, ev_x is the "evaluate at x" map for both functions and vector fields.

Theorem 6.4 (Weak Characterisations of Measurability, Integrability, and Boundedness of Smooth Time-Varying Vector Fields). *Let* M *be a smooth manifold, let* $\mathbb{T} \subseteq \mathbb{R}$ *be a time-domain, and let* $X\colon \mathbb{T} \times \mathsf{M} \to \mathsf{TM}$ *have the property that* X_t *is a smooth vector field for each* $t \in \mathbb{T}$. *Then the following four statements are equivalent:*

(i) $t \mapsto X_t$ *is measurable;*
(ii) $t \mapsto \mathscr{L}_{X_t}f$ *is measurable for every* $f \in \mathrm{C}^\infty(\mathsf{M})$*;*
(iii) $t \mapsto \mathrm{ev}_x \circ X_t$ *is measurable for every* $x \in \mathsf{M}$*;*
(iv) $t \mapsto \mathrm{ev}_x \circ \mathscr{L}_{X_t}f$ *is measurable for every* $f \in \mathrm{C}^\infty(\mathsf{M})$ *and every* $x \in \mathsf{M}$,
the following two statements are equivalent:
(v) $t \mapsto X_t$ *is locally Bochner integrable;*
(vi) $t \mapsto \mathscr{L}_{X_t}f$ *is locally Bochner integrable for every* $f \in \mathrm{C}^\infty(\mathsf{M})$,

and the following two statements are equivalent:
(vii) $t \mapsto X_t$ *is locally essentially von Neumann bounded;*
(viii) $t \mapsto \mathscr{L}_{X_t} f$ *is locally essentially von Neumann bounded for every* $f \in C^\infty(\mathsf{M})$.

Proof: This follows from Theorem 6.3, along with Corollary 3.6. □

Let us now discuss flows of vector fields from $\mathrm{LI}\Gamma^\infty(\mathbb{T};\mathsf{TM})$. To do so, let us provide the definition of the usual attribute of integral curves, but on manifolds.

Definition 6.5 (Locally Absolutely Continuous). Let M be a smooth manifold and let $\mathbb{T} \subseteq \mathbb{R}$ be an interval.

(i) A function $f\colon [a,b] \to \mathbb{R}$ is ***absolutely continuous*** if there exists $g \in \mathrm{L}^1([\mathrm{a},\mathrm{b}];\mathbb{R})$ such that

$$f(t) = f(a) + \int_a^t g(\tau)\,\mathrm{d}\tau, \qquad t \in [a,b].$$

(ii) A function $f\colon \mathbb{T} \to \mathbb{R}$ is ***locally absolutely continuous*** if $f|\mathbb{T}'$ is absolutely continuous for every compact subinterval $\mathbb{T}' \subseteq \mathbb{T}$.

(iii) A curve $\gamma\colon \mathbb{T} \to \mathsf{M}$ is ***locally absolutely continuous*** if $\phi \circ \gamma$ is locally absolutely continuous for every $\phi \in C^\infty(\mathsf{M})$. ○

One easily verifies that a curve is locally absolutely continuous according to our definition if and only if its local representative is locally absolutely continuous in any coordinate chart.

We then have the following existence, uniqueness, and regularity result for locally integrally bounded vector fields. In the statement of the result, we use the notation

$$|a,b| = \begin{cases} [a,b], & a \le b, \\ [b,a], & b < a. \end{cases}$$

In the following result, we do not provide the comprehensive list of properties of the flow, but only those required to make sense of its regularity with respect to initial conditions.

Theorem 6.6 (Flows of Vector Fields from $\mathrm{LI}\Gamma^\infty(\mathbb{T};\mathsf{TM})$). *Let* M *be a smooth manifold, let* $\mathbb{T}$ *be an interval, and let* $X \in \mathrm{LI}\Gamma^\infty(\mathbb{T};\mathsf{TM})$. *Then there exist a subset* $D_X \subseteq \mathbb{T}\times\mathbb{T}\times\mathsf{M}$ *and a map* $\Phi^X\colon D_X \to \mathsf{M}$ *with the following properties for each* $(t_0,x_0) \in \mathbb{T}\times\mathsf{M}$:

(i) the set

$$\mathbb{T}_X(t_0,x_0) = \{t \in \mathbb{T} \mid (t,t_0,x_0) \in D_X\}$$

is an interval;

(ii) there exists a locally absolutely continuous curve $t \mapsto \xi(t)$ *satisfying*

$$\xi'(t) = X(t,\xi(t)), \quad \xi(t_0) = x_0,$$

for almost all $t \in |t_0,t_1|$ *if and only if* $t_1 \in \mathbb{T}_X(t_0,x_0)$;

(iii) $\frac{\mathrm{d}}{\mathrm{d}t}\Phi^X(t,t_0,x_0) = X(t,\Phi^X(t,t_0,x_0))$ *for almost all* $t \in \mathbb{T}_X(t_0,x_0)$;

(iv) *for each* $t \in \mathbb{T}$ *for which* $(t, t_0, x_0) \in D_X$*, there exists a neighbourhood* $\mathcal{U}$ *of* x_0 *such that the mapping* $x \mapsto \Phi^X(t, t_0, x)$ *is defined and of class* C^∞ *on* $\mathcal{U}$.

Proof: We observe that the requirement that $X \in \mathrm{LI}\Gamma^\infty(\mathbb{T}; \mathsf{TM})$ implies that, in any coordinate chart, the components of X and their derivatives are all bounded by a locally integrable function. This, in particular, implies that, in any coordinate chart for M, the ordinary differential equation associated with the vector field X satisfies the usual conditions for existence and uniqueness of solutions as per, for example, [16, Theorem 54]. Of course, the differential equation satisfies conditions much stronger than this, and we shall see how to use these in our argument below.

The first three assertions are now part of the standard existence theorem for solutions of ordinary differential equations, along with the usual Zorn's Lemma argument for the existence of a maximal interval on which integral curves is defined.

In the sequel we denote $\Phi^X_{t,t_0}(x) = \Phi^X(t, t_0, x_0)$.

For the fourth assertion we first make some constructions with vector fields on jet bundles, more or less following [13, Sect. 4.4]. We let $\mathsf{M}^2 = \mathsf{M} \times \mathsf{M}$ and we consider M^2 as a fibred manifold, indeed a trivial fibre bundle, over M by $\mathrm{pr}_1 : \mathsf{M}^2 \to \mathsf{M}$, i.e., by projection onto the first factor. A section of this fibred manifold is naturally identified with a smooth map $\Phi: \mathsf{M} \to \mathsf{M}$ by $x \mapsto (x, \Phi(x))$. We introduce the following notation:

1. $\mathsf{J}^m \mathrm{pr}_1$: the bundle of m-jets of sections of the fibred manifold $\mathrm{pr}_1 : \mathsf{M}^2 \to \mathsf{M}$;
2. $\mathsf{V}\,\mathrm{pr}_{1,m}$: the vertical bundle of the fibred manifold $\mathrm{pr}_{1,m} : \mathsf{J}^m \mathrm{pr}_1 \to \mathsf{M}$;
3. $\mathsf{V}\,\mathrm{pr}_1$: the vertical bundle of the fibred manifold $\mathrm{pr}_1 : \mathsf{M}^2 \to \mathsf{M}$;
4. ν: the projection $\mathrm{pr}_1 \circ (\pi_{\mathsf{TM}^2} | \mathsf{V}\,\mathrm{pr}_1)$;
5. $\mathsf{J}^m \nu$: the bundle of m-jets of sections of the fibred manifold $\nu : \mathsf{V}\,\mathrm{pr}_1 \to \mathsf{M}$.

With this notation, we have the following lemma.

Lemma 1 *There is a canonical diffeomorphism* $\alpha_m : \mathsf{J}^m \nu \to \mathsf{V}\,\mathrm{pr}_{1,m}$.

Proof: We describe the diffeomorphism, and then note that the verification that it is, in fact, a diffeomorphism is a fact easily checked in jet bundle coordinates.

Let $I \subseteq \mathbb{R}$ be an interval with $0 \in \mathrm{int}(I)$ and consider a smooth map $\phi: I \times \mathsf{M} \to \mathsf{M} \times \mathsf{M}$ of the form $\phi(t, x) = (x, \phi_1(t, x))$ for a smooth map ϕ_1. We let $\phi_t(x) = \phi^x(t) = \phi(t, x)$. We then have maps

$$\begin{aligned} j^x_m\phi : I &\to \mathsf{J}^m \mathrm{pr}_1 \\ t &\mapsto j_m\phi_t(x) \end{aligned}$$

and

$$\begin{aligned} \phi' : \mathsf{M} &\to \mathsf{V}\,\mathrm{pr}_1 \\ x &\mapsto \frac{\mathrm{d}}{\mathrm{d}t}\Big|_{t=0} \phi^x(t). \end{aligned}$$

Note that the curve $j^x_m\phi$ is a curve in the fibre of $\mathrm{pr}_{1,m} : \mathsf{J}^m \mathrm{pr}_1 \to \mathsf{M}$. Thus we can sensibly define α_m by

$$\alpha_m(j_m\phi'(x)) = \frac{\mathrm{d}}{\mathrm{d}t}\Big|_{t=0} j^x_m\phi(t).$$

In jet bundle coordinates, one can check that α_m has the local representative

$$((\boldsymbol{x}_1, (\boldsymbol{x}_2, \boldsymbol{A}_0)), (\boldsymbol{B}_1, \boldsymbol{A}_1, \dots, \boldsymbol{B}_m, \boldsymbol{A}_m)) \\ \mapsto ((\boldsymbol{x}_1, (\boldsymbol{x}_2, \boldsymbol{B}_1, \dots, \boldsymbol{B}_m)), (\boldsymbol{A}_0, \boldsymbol{A}_1, \dots, \boldsymbol{A}_m)),$$

showing that α_m is indeed a diffeomorphism. ▽

Given a smooth vector field Y on M, we define a vector field $\tilde{Y}$ on M^2 by $\tilde{Y}(x_1, x_2) = (0_{x_1}, Y(x_2))$. Note that we have the following commutative diagram

$$\begin{array}{ccc} \mathsf{M}^2 & \xrightarrow{\tilde{Y}} & \mathsf{V}\,\mathrm{pr}_1 \\ {\scriptstyle \mathrm{pr}_1}\downarrow & & \downarrow{\scriptstyle \nu} \\ \mathsf{M} & = & \mathsf{M} \end{array}$$

giving $\tilde{Y}$ as a morphism of fibred manifolds. It is thus a candidate to have its m-jet taken, giving a morphism of fibred manifolds $j_m\tilde{Y}\colon \mathsf{J}^m\,\mathrm{pr}_1 \to \mathsf{J}^m\nu$. By the lemma, $\alpha_m \circ j_m\tilde{Y}$ is a vertical vector field on $\mathsf{J}^m\,\mathrm{pr}_1$ that we denote by $\nu_m Y$, the ***mth vertical prolongation*** of Y. Let us verify that this is a vector field. First of all, for a section $\tilde{\Phi}$ of pr_1 given by $x \mapsto (x, \Phi(x))$, note that $\tilde{Y} \circ \tilde{\Phi}(x) = (0_x, Y(\Phi(x)))$, and so $j_m(\tilde{Y} \circ \tilde{\Phi})(x)$ is vertical. By the notation from the proof of the lemma, we can write $j_m(\tilde{Y} \circ \tilde{\Phi})(x) = j_m\phi'(x)$ for some suitable map ϕ as in the lemma. We then have

$$\alpha_m \circ j_m(\tilde{Y} \circ \tilde{\Phi})(x) = \alpha_m(j_m\phi'(x)) \in \mathsf{V}_{j_m\tilde{\Phi}(x)}\,\mathrm{pr}_{1,m}\,.$$

Therefore,

$$\pi_{\mathsf{TJ}^m\,\mathrm{pr}_1}(\alpha_m \circ j_m\tilde{Y}(j_m\tilde{\Phi}(x))) = \pi_{\mathsf{TJ}^m\,\mathrm{pr}_1}(\alpha_m \circ j_m(\tilde{Y} \circ \tilde{\Phi})(x)) = j_m\tilde{\Phi}(x).$$

Note that, since $\mathsf{J}^m\,\mathrm{pr}_1$ is naturally identified with $\mathsf{J}^m(\mathsf{M};\mathsf{M})$ via the identification

$$j_m\tilde{\Phi}(x) \mapsto j_m\Phi(x),$$

if $\tilde{\Phi}(x) = (x, \Phi(x))$, we can as well think of $\nu_m Y$ as being a vector field on the latter space. Sorting through all the definitions gives the form of $\nu_m Y$ in coordinates as

$$((\boldsymbol{x}_1, \boldsymbol{x}_2), \boldsymbol{A}_1, \dots, \boldsymbol{A}_m) \mapsto (((\boldsymbol{x}_1, \boldsymbol{x}_2), \boldsymbol{A}_1, \dots, \boldsymbol{A}_m), \boldsymbol{0}, \boldsymbol{Y}, \boldsymbol{DY}, \dots, \boldsymbol{D}^m\boldsymbol{Y}). \quad (6.1)$$

We now apply the above constructions, for each fixed $t \in \mathbb{T}$, to get the vector field $\nu_m X_t$, and so the time-varying vector field $\nu_m X$ defined by $\nu_m X(t, j_m\Phi(x)) = \nu_m X_t(j_m\Phi(x))$ on $\mathsf{J}^m(\mathsf{M};\mathsf{M})$. The definition of $\mathrm{LI}\Gamma^1(\mathbb{T};\mathsf{TM})$, along with the coordinate formula (6.1), shows that $\nu_m X$ satisfies the standard conditions for existence and uniqueness of integral curves, and so its flow depends continuously on initial condition [16, Theorem 55].

The fourth part of the theorem, therefore, will follow if we can show that

1. for each $m \in \mathbb{Z}_{\geq 0}$, the flow of $\nu_m X$ depends on the initial condition in M in a C^m way,
2. $\Phi^{\nu_m X}_{t,t_0}(j_m \Phi^X_{t_0,t_0}(x_0)) = j_m \Phi^X_{t,t_0}(j_m \Phi^X_{t_0,t_0}(x_0))$, and
3. if $\{t\} \times \{t_0\} \times \mathcal{U} \subseteq D_X$, then $\{t\} \times \{t_0\} \times \mathrm{pr}^{-1}_{1,m}(\mathcal{U}) \subseteq D_{\nu_m X}$.

We ask for property 3 to ensure that the domain of differentiability does not get too small as the order of the derivatives gets large.

To prove these assertions, it suffices to work locally. According to (6.1), we have the time-dependent differential equation defined on

$$\mathcal{U} \times \mathrm{L}(\mathbb{R}^n; \mathbb{R}^n) \times \cdots \times \mathrm{L}^m_{\mathrm{sym}}(\mathbb{R}^n; \mathbb{R}^n),$$

where $\mathcal{U}$ is an open subset of $\mathbb{R}^n$, and given by

$$\begin{aligned}
\dot{\gamma}(t) &= \boldsymbol{X}(t, \gamma(t)),\\
\dot{\boldsymbol{A}}_1(t) &= \boldsymbol{D}\boldsymbol{X}(t, \gamma(t)),\\
\dot{\boldsymbol{A}}_2(t) &= \boldsymbol{D}^2\boldsymbol{X}(t, \gamma(t)),\\
&\ \ \vdots\\
\dot{\boldsymbol{A}}_m(t) &= \boldsymbol{D}^m\boldsymbol{X}(t, \gamma(t)),
\end{aligned}$$

$(t, \boldsymbol{x}) \mapsto (\boldsymbol{x}, \boldsymbol{X}(t, \boldsymbol{x}))$ being the local representative of X. The initial conditions of interest for the vector field $\nu_m X$ are of the form $j_m \Phi^X_{t_0,t_0}(x)$. In coordinates, keeping in mind that $\Phi^X_{t_0,t_0} = \mathrm{id}_{\mathsf{M}}$, this gives

$$\gamma(t_0) = \boldsymbol{x}_0,\ \boldsymbol{A}_1(t_0) = \boldsymbol{I}_n,\ \boldsymbol{A}_j(t_0) = \boldsymbol{0}, \qquad j \geq 2. \tag{6.2}$$

Let us denote by $t \mapsto \gamma(t, t_0, \boldsymbol{x})$ and $t \mapsto \boldsymbol{A}_j(t, t_0, \boldsymbol{x})$, $j \in \{1, \ldots, m\}$, the solutions of the differential equations above with these initial conditions.

We will show that assertions 1–3 hold by induction on m. In doing this, we will need to understand how differential equations depending differentiably on state also have solutions depending differentiably on initial condition. Such a result is not readily found in the textbook literature, as this latter is typically concerned with continuous dependence on initial conditions for cases with measurable time dependence, and on differentiable dependence when the dependence on time is also differentiable. However, the general case (much more general than we need here) is worked out in [15].

For $m = 0$, the assertions are simply the result of the usual continuous dependence on initial conditions, e.g., [16, Theorem 55]. Let us consider the case $m = 1$. In this case, the properties of $\mathrm{LI}\Gamma^\infty(\mathbb{T}; \mathsf{TM})$ ensure that the hypotheses required to apply Theorem 2.1 of [15] hold for the differential equation

$$\begin{aligned}
\dot{\gamma}(t) &= \boldsymbol{X}(t, \gamma(t)),\\
\dot{\boldsymbol{A}}_1(t) &= \boldsymbol{D}\boldsymbol{X}(t, \gamma(t)).
\end{aligned}$$

This allows us to conclude that $\boldsymbol{x} \mapsto \boldsymbol{\gamma}(t, t_0, \boldsymbol{x})$ is of class C^1. This establishes the assertion 1 in this case. Therefore, on a suitable domain, $j_1\Phi^X_{t,t_0}$ is well defined. In coordinates the map $j_1\Phi^X_{t,t_0} \colon \mathsf{J}^1(\mathsf{M};\mathsf{M}) \to \mathsf{J}^1(\mathsf{M};\mathsf{M})$ is given by

$$(\boldsymbol{x}, \boldsymbol{y}, \boldsymbol{B}_1) \mapsto (\boldsymbol{x}, \boldsymbol{\gamma}(t, t_0, \boldsymbol{x}), \boldsymbol{D}_3\boldsymbol{\gamma}(t, t_0, \boldsymbol{x}) \circ \boldsymbol{B}_1), \tag{6.3}$$

this by the Chain Rule. We have

$$\frac{\mathrm{d}}{\mathrm{d}t}\boldsymbol{D}_3\boldsymbol{\gamma}(t, t_0, \boldsymbol{x}) = \boldsymbol{D}_3(\tfrac{\mathrm{d}}{\mathrm{d}t}\boldsymbol{\gamma}(t, t_0, \boldsymbol{x})) = \boldsymbol{D}\boldsymbol{X}(t, \boldsymbol{\gamma}(t, t_0, \boldsymbol{x})),$$

the swapping of the time and spatial derivatives being valid by [15, Corollary 2.2]. Combining this with (6.3) and the initial conditions (6.2) shows that assertion 2 holds for $m = 1$. Moreover, since $\boldsymbol{A}_1(t, t_0, \boldsymbol{x})$ is obtained by merely integrating a continuous function of t from t_0 to t, we also conclude that assertion 3 holds.

Now suppose that assertions 1–3 hold for m. Again, the properties of $\mathrm{LI}\Gamma^\infty(\mathbb{T};\mathsf{TM})$ imply that the hypotheses of Theorem 2.1 of [15] hold, and so solutions of the differential equation

$$\begin{aligned}
\dot{\boldsymbol{\gamma}}(t) &= \boldsymbol{X}(t, \boldsymbol{\gamma}(t)),\\
\dot{\boldsymbol{A}}_1(t) &= \boldsymbol{D}\boldsymbol{X}(t, \boldsymbol{\gamma}(t)),\\
\dot{\boldsymbol{A}}_2(t) &= \boldsymbol{D}^2\boldsymbol{X}(t, \boldsymbol{\gamma}(t)),\\
&\ \ \vdots\\
\dot{\boldsymbol{A}}_m(t) &= \boldsymbol{D}^m\boldsymbol{X}(t, \boldsymbol{\gamma}(t))
\end{aligned}$$

depend continuously differentiably on initial condition. By the induction hypothesis applied to the assertion 2, this means that

$$(t, x) \mapsto \Phi^{\nu_m X}_{t,t_0}(j_m\Phi^X_{t_0,t_0}(x)) = j_m\Phi^X_{t,t_0}(x)$$

depends continuously differentiably on x, and so we conclude that $(t, x) \mapsto \Phi^X_{t,t_0}(x)$ depends on x in a C^{m+1} manner. This establishes assertion 1 for $m + 1$. After an application of the Chain Rule for high-order derivatives (see [1, Supplement 2.4A]) we can, admittedly after just a few moments thought, see that the local representative of $j_{m+1}\Phi^X_{t,t_0}(j_{m+1}\Phi^X_{t_0,t_0}(x))$ is

$$(\boldsymbol{x}, \boldsymbol{\gamma}(t, t_0, \boldsymbol{x}), \boldsymbol{D}_3\boldsymbol{\gamma}(t, t_0, \boldsymbol{x}), \dots, \boldsymbol{D}_3^{m+1}\boldsymbol{\gamma}(t, t_0, \boldsymbol{x})),$$

keeping in mind the initial conditions (6.2) in coordinates.

By the induction hypothesis,

$$\frac{\mathrm{d}}{\mathrm{d}t}\boldsymbol{D}_3^j\boldsymbol{\gamma}(t) = \boldsymbol{D}^j\boldsymbol{X}(t, \boldsymbol{\gamma}(t, t_0, \boldsymbol{x})), \qquad j \in \{1, \dots, m\}.$$

Using Corollary 2.2 of [15] we compute

$$\frac{\mathrm{d}}{\mathrm{d}t} \boldsymbol{D}_3^{m+1} \boldsymbol{\gamma}(t, t_0, \boldsymbol{x}) = \boldsymbol{D}(\tfrac{\mathrm{d}}{\mathrm{d}t} \boldsymbol{D}_3^m \boldsymbol{\gamma}(t, t_0, \boldsymbol{x})) = \boldsymbol{D}^{m+1} \boldsymbol{X}(t, \boldsymbol{\gamma}(t, t_0, \boldsymbol{x})),$$

giving assertion 2 for $m + 1$. Finally, by the induction hypothesis and since $\boldsymbol{A}_{m+1}(t, t_0, \boldsymbol{x})$ is obtained by simple integration from t_0 to t, we conclude that assertion 3 holds for $m + 1$. □

6.2 The Finitely Differentiable or Lipschitz Case

The requirement that the flow depends smoothly on initial conditions is not always essential, even when the vector field itself depends smoothly on the state. In such cases as this, one may want to consider classes of vector fields characterised by one of the weaker topologies described in Sect. 3.4. Let us see how to do this. In this section, so as to be consistent with our definition of Lipschitz norms in Sect. 3.5, we suppose that the affine connection ∇ on M is the Levi–Civita connection for the Riemannian metric $\mathbb{G}$ and that the vector bundle connection ∇^0 in E is $\mathbb{G}_0$-orthogonal.

Definition 6.7 (Finitely Differentiable or Lipschitz Carathéodory Section). Let $\pi\colon \mathsf{E} \to \mathsf{M}$ be a smooth vector bundle and let $\mathbb{T} \subseteq \mathbb{R}$ be an interval. Let $m \in \mathbb{Z}_{\geq 0}$ and let $m' \in \{0, \mathrm{lip}\}$. A ***Carathéodory section of class $\mathrm{C}^{m+m'}$*** of E is a map $\xi\colon \mathbb{T}\times\mathsf{M} \to \mathsf{E}$ with the following properties:

(i) $\xi(t, x) \in \mathsf{E}_x$ for each $(t, x) \in \mathbb{T} \times \mathsf{M}$;
(ii) for each $t \in \mathbb{T}$, the map $\xi_t\colon \mathsf{M} \to \mathsf{E}$ defined by $\xi_t(x) = \xi(t, x)$ is of class $\mathrm{C}^{m+m'}$;
(iii) for each $x \in \mathsf{M}$, the map $\xi^x\colon \mathbb{T} \to \mathsf{E}$ defined by $\xi^x(t) = \xi(t, x)$ is Lebesgue measurable.

We shall call $\mathbb{T}$ the ***time-domain*** for the section. By $\mathrm{CF}\Gamma^{m+m'}(\mathbb{T}; \mathsf{E})$ we denote the set of Carathéodory sections of class $\mathrm{C}^{m+m'}$ of E. ○

Now we put some conditions on the time dependence of the derivatives of the section.

Definition 6.8 (Locally Integrally $\mathrm{C}^{m+m'}$-Bounded and Locally Essentially $\mathrm{C}^{m+m'}$-Bounded Sections). Let $\pi\colon \mathsf{E} \to \mathsf{M}$ be a smooth vector bundle and let $\mathbb{T} \subseteq \mathbb{R}$ be an interval. Let $m \in \mathbb{Z}_{\geq 0}$ and let $m' \in \{0, \mathrm{lip}\}$. A Carathéodory section $\xi\colon \mathbb{T} \times \mathsf{M} \to \mathsf{E}$ of class $\mathrm{C}^{m+m'}$ is

(i) ***locally integrally $\mathrm{C}^{m+m'}$-bounded*** if:
  (a) $m' = 0$: for every compact set $K \subseteq \mathsf{M}$, there exists $g \in \mathrm{L}^1_{\mathrm{loc}}(\mathbb{T}; \mathbb{R}_{\geq 0})$ such that

$$\|j_m\xi_t(x)\|_{\overline{\mathbb{G}}_m} \leq g(t), \qquad (t, x) \in \mathbb{T} \times K;$$

  (b) $m' = \mathrm{lip}$: for every compact set $K \subseteq \mathsf{M}$, there exists $g \in \mathrm{L}^1_{\mathrm{loc}}(\mathbb{T}; \mathbb{R}_{\geq 0})$ such that

$$\mathrm{dil}\, j_m\xi_t(x), \|j_m\xi_t(x)\|_{\overline{\mathbb{G}}_m} \leq g(t), \qquad (t, x) \in \mathbb{T} \times K,$$

and is

(ii) ***locally essentially*** $\mathrm{C}^{m+m'}$***-bounded*** if:

(a) $m' = 0$: for every compact set $K \subseteq \mathsf{M}$, there exists $g \in \mathrm{L}^\infty_{\mathrm{loc}}(\mathbb{T}; \mathbb{R}_{\geq 0})$ such that

$$\|j_m\xi_t(x)\|_{\overline{\mathbb{G}}_m} \leq g(t), \qquad (t, x) \in \mathbb{T} \times K;$$

(b) $m' = \mathrm{lip}$: for every compact set $K \subseteq \mathsf{M}$, there exists $g \in \mathrm{L}^\infty_{\mathrm{loc}}(\mathbb{T}; \mathbb{R}_{\geq 0})$ such that

$$\mathrm{dil}\, j_m\xi_t(x), \|j_m\xi_t(x)\|_{\overline{\mathbb{G}}_m} \leq g(t), \qquad (t, x) \in \mathbb{T} \times K.$$

The set of locally integrally $\mathrm{C}^{m+m'}$-bounded sections of E with time-domain $\mathbb{T}$ is denoted by $\mathrm{LI}\Gamma^{m+m'}(\mathbb{T}, \mathsf{E})$ and the set of locally essentially $\mathrm{C}^{m+m'}$-bounded sections of E with time-domain $\mathbb{T}$ is denoted by $\mathrm{LB}\Gamma^{m+m'}(\mathbb{T}; \mathsf{E})$. ○

Theorem 6.9 (Topological Characterisation of Finitely Differentiable or Lipschitz Carathéodory Sections). *Let* $\pi\colon \mathsf{E} \to \mathsf{M}$ *be a smooth vector bundle and let* $\mathbb{T} \subseteq \mathbb{R}$ *be an interval. Let* $m \in \mathbb{Z}_{\geq 0}$ *and let* $m' \in \{0, \mathrm{lip}\}$*. For a map* $\xi\colon \mathbb{T} \times \mathsf{M} \to \mathsf{E}$ *satisfying* $\xi(t, x) \in \mathsf{E}_x$ *for each* $(t, x) \in \mathbb{T} \times \mathsf{M}$*, the following two statements are equivalent:*

(i) $\xi \in \mathrm{CF}\Gamma^{m+m'}(\mathbb{T}; \mathsf{E})$*;*

(ii) the map $\mathbb{T} \ni t \mapsto \xi_t \in \Gamma^{m+m'}(\mathsf{E})$ *is measurable,*

the following two statements are equivalent:

(iii) $\xi \in \mathrm{LI}\Gamma^{m+m'}(\mathbb{T}; \mathsf{E})$*;*

(iv) the map $\mathbb{T} \ni t \mapsto \xi_t \in \Gamma^{m+m'}(\mathsf{E})$ *is measurable and locally Bochner integrable,*

and the following two statements are equivalent:

(v) $\xi \in \mathrm{LB}\Gamma^{m+m'}(\mathbb{T}; \mathsf{E})$*;*

(vi) the map $\mathbb{T} \ni t \mapsto \xi_t \in \Gamma^{m+m'}(\mathsf{E})$ *is measurable and locally essentially von Neumann bounded.*

Proof: (i) $\Longleftrightarrow$ (ii) For $x \in \mathsf{M}$ and $\alpha_x \in \mathsf{E}^*_x$, define $\mathrm{ev}_{\alpha_x}\colon \Gamma^{m+m'}(\mathsf{E}) \to \mathbb{R}$ by $\mathrm{ev}_{\alpha_x}(\xi) = \langle \alpha_x; \xi(x)\rangle$. It is easy to show that ev_{α_x} is continuous and that the set of continuous functionals ev_{α_x}, $\alpha_x \in \mathsf{E}^*_x$, is point separating. Since $\Gamma^{m+m'}(\mathsf{E})$ is a Suslin space (properties CO^m-6 and $\mathrm{CO}^{m+\mathrm{lip}}$-6), this part of the theorem follows in the same manner as the corresponding part of Theorem 6.3.

(iii) $\Longleftrightarrow$ (iv) Since $\Gamma^{m+m'}(\mathsf{E})$ is complete and separable (by properties CO^m-2 and CO^m-4, and $\mathrm{CO}^{m+\mathrm{lip}}$-2 and $\mathrm{CO}^{m+\mathrm{lip}}$-4), the arguments from the corresponding part of Theorem 6.3 apply here, taking note of the definition of the seminorms $p^{\mathrm{lip}}_K(\xi)$ in case $m' = \mathrm{lip}$.

(v) $\Longleftrightarrow$ (vi) We recall our discussion of von Neumann bounded sets in locally convex topological vector spaces preceding Lemma 3.1 above. With this in mind and using Lemma 4.3, this part of the proposition follows immediately. □

Note that Theorem 6.9 applies, in particular, to vector fields and functions, giving the classes $\mathrm{CF}^{m+m'}(\mathbb{T}; \mathsf{M})$, $\mathrm{LIC}^{m+m'}(\mathbb{T}; \mathsf{M})$, and $\mathrm{LBC}^{m+m'}(\mathbb{T}; \mathsf{M})$ of functions, and the classes $\mathrm{CF}\Gamma^{m+m'}(\mathbb{T}; \mathsf{TM})$, $\mathrm{LI}\Gamma^{m+m'}(\mathbb{T}; \mathsf{TM})$, and $\mathrm{LB}\Gamma^{m+m'}(\mathbb{T}; \mathsf{TM})$ of vector fields. Noting that we have the alternative weak-$\mathscr{L}$ characterisation of the $\mathrm{CO}^{m+m'}$-topology, we can summarise the various sorts of measurability, integrability, and boundedness for smooth time-varying vector fields as follows. In the statement of the result, ev_x is the "evaluate at x" map for both functions and vector fields.

Theorem 6.10 (Weak Characterisations of Measurability, Integrability, and Boundedness of Finitely Differentiable or Lipschitz Time-Varying Vector Fields). *Let* M *be a smooth manifold, let* $\mathbb{T} \subseteq \mathbb{R}$ *be a time-domain, let* $m \in \mathbb{Z}_{\geq 0}$*, let* $m' \in \{0, \mathrm{lip}\}$*, and let* $X : \mathbb{T} \times \mathsf{M} \to \mathsf{TM}$ *have the property that* X_t *is a vector field of class* $C^{m+m'}$ *for each* $t \in \mathbb{T}$*. Then the following four statements are equivalent:*

(i) $t \mapsto X_t$ *is measurable;*
(ii) $t \mapsto \mathscr{L}_{X_t} f$ *is measurable for every* $f \in \mathrm{C}^\infty(\mathsf{M})$*;*
(iii) $t \mapsto \mathrm{ev}_x \circ X_t$ *is measurable for every* $x \in \mathsf{M}$*;*
(iv) $t \mapsto \mathrm{ev}_x \circ \mathscr{L}_{X_t} f$ *is measurable for every* $f \in \mathrm{C}^\infty(\mathsf{M})$ *and every* $x \in \mathsf{M}$*,*

the following two statements are equivalent:

(v) $t \mapsto X_t$ *is locally Bochner integrable;*
(vi) $t \mapsto \mathscr{L}_{X_t} f$ *is locally Bochner integrable for every* $f \in \mathrm{C}^\infty(\mathsf{M})$*,*

and the following two statements are equivalent:

(vii) $t \mapsto X_t$ *is locally essentially von Neumann bounded;*
(viii) $t \mapsto \mathscr{L}_{X_t} f$ *is locally essentially von Neumann bounded for every* $f \in \mathrm{C}^\infty(\mathsf{M})$*.*

Proof: This follows from Theorem 6.9, along with Corollaries 3.9 and 3.15. □

It is also possible to state an existence, uniqueness, and regularity theorem for flows of vector fields that depend on state in a finitely differentiable or Lipschitz manner.

Theorem 6.11 (Flows of Vector Fields from $\mathrm{LI}\Gamma^{m+m'}(\mathbb{T}; \mathsf{TM})$). *Let* M *be a smooth manifold, let* $\mathbb{T}$ *be an interval, let* $m \in \mathbb{Z}_{\geq 0}$*, and let* $X \in \mathrm{LI}\Gamma^{m+\mathrm{lip}}(\mathbb{T}; \mathsf{TM})$*. Then there exist a subset* $D_X \subseteq \mathbb{T} \times \mathbb{T} \times \mathsf{M}$ *and a map* $\Phi^X : D_X \to \mathsf{M}$ *with the following properties for each* $(t_0, x_0) \in \mathbb{T} \times \mathsf{M}$*:*

(i) the set

$$\mathbb{T}_X(t_0, x_0) = \{t \in \mathbb{T} \mid (t, t_0, x_0) \in D_X\}$$

is an interval;

(ii) there exists a locally absolutely continuous curve $t \mapsto \xi(t)$ *satisfying*

$$\xi'(t) = X(t, \xi(t)), \quad \xi(t_0) = x_0,$$

for almost all $t \in |t_0, t_1|$ *if and only if* $t_1 \in \mathbb{T}_X(t_0, x_0)$*;*

(iii) $\frac{\mathrm{d}}{\mathrm{d}t} \Phi^X(t, t_0, x_0) = X(t, \Phi^X(t, t_0, x_0))$ *for almost all* $t \in \mathbb{T}_X(t_0, x_0)$*;*

(iv) for each $t \in \mathbb{T}$ *for which* $(t, t_0, x_0) \in D_X$*, there exists a neighbourhood* $\mathcal{U}$ *of* x_0 *such that the mapping* $x \mapsto \Phi^X(t, t_0, x)$ *is defined and of class* C^m *on* $\mathcal{U}$*.*

Proof: The proof here is by truncation of the proof of Theorem 6.6 from "∞" to "m". □

6.3 The Holomorphic Case

While we are not per se interested in time-varying holomorphic vector fields, our understanding of time-varying real analytic vector fields—in which we are most definitely interested—is connected with an understanding of the holomorphic case, cf. Theorem 6.25.

We begin with definitions that are similar to the smooth case, but which rely on the holomorphic topologies introduced in Sect. 4.1. We will consider an holomorphic vector bundle $\pi\colon \mathsf{E} \to \mathsf{M}$ with an Hermitian fibre metric $\mathbb{G}$. This defines the seminorms p_K^{hol}, $K \subseteq \mathsf{M}$ compact, describing the CO^{hol}-topology for $\Gamma^{\text{hol}}(\mathsf{E})$ as in Sect. 4.1.

Let us get started with the definitions.

Definition 6.12 (Holomorphic Carathéodory Section). Let $\pi\colon \mathsf{E} \to \mathsf{M}$ be an holomorphic vector bundle and let $\mathbb{T} \subseteq \mathbb{R}$ be an interval. A ***Carathéodory section of class C^{hol}*** of E is a map $\xi\colon \mathbb{T} \times \mathsf{M} \to \mathsf{E}$ with the following properties:

(i) $\xi(t, z) \in \mathsf{E}_z$ for each $(t, z) \in \mathbb{T} \times \mathsf{M}$;

(ii) for each $t \in \mathbb{T}$, the map $\xi_t\colon \mathsf{M} \to \mathsf{E}$ defined by $\xi_t(z) = \xi(t, z)$ is of class C^{hol};

(iii) for each $z \in \mathsf{M}$, the map $\xi^z\colon \mathbb{T} \to \mathsf{E}$ defined by $\xi^z(t) = \xi(t, z)$ is Lebesgue measurable.

We shall call $\mathbb{T}$ the ***time-domain*** for the section. By CF$\Gamma^{\text{hol}}(\mathbb{T}; \mathsf{E})$ we denote the set of Carathéodory sections of class C^{hol} of E. ∘

The associated notions for time-dependent sections compatible with the CO^{hol}-topology are as follows.

Definition 6.13 (Locally Integrally C^{hol}-Bounded and Locally Essentially C^{hol}-Bounded Sections). Let $\pi\colon \mathsf{E} \to \mathsf{M}$ be an holomorphic vector bundle and let $\mathbb{T} \subseteq \mathbb{R}$ be an interval. A Carathéodory section $\xi\colon \mathbb{T} \times \mathsf{M} \to \mathsf{E}$ of class C^{hol} is

(i) ***locally integrally C^{hol}-bounded*** if, for every compact set $K \subseteq \mathsf{M}$, there exists $g \in \mathrm{L}^1_{\text{loc}}(\mathbb{T}; \mathbb{R}_{\geq 0})$ such that

$$\|\xi(t, z)\|_{\mathbb{G}} \leq g(t), \qquad (t, z) \in \mathbb{T} \times K$$

and is

(ii) ***locally essentially C^{hol}-bounded*** if, for every compact set $K \subseteq \mathsf{M}$, there exists $g \in \mathrm{L}^\infty_{\text{loc}}(\mathbb{T}; \mathbb{R}_{\geq 0})$ such that

$$\|\xi(t, z)\|_{\mathbb{G}} \leq g(t), \qquad (t, z) \in \mathbb{T} \times K.$$

The set of locally integrally C^{hol}-bounded sections of E with time-domain $\mathbb{T}$ is denoted by LI$\Gamma^{\text{hol}}(\mathbb{T}, \mathsf{E})$ and the set of locally essentially C^{hol}-bounded sections of E with time-domain $\mathbb{T}$ is denoted by LB$\Gamma^{\text{hol}}(\mathbb{T}; \mathsf{E})$. ∘

As with smooth sections, the preceding definitions admit topological characterisations, now using the CO^{hol}-topology for $\Gamma^{\text{hol}}(\mathsf{E})$.

Theorem 6.14 (Topological Characterisation of Holomorphic Carathéodory Sections). *Let $\pi\colon \mathsf{E} \to \mathsf{M}$ be an holomorphic vector bundle and let $\mathbb{T} \subseteq \mathbb{R}$ be an interval. For a map $\xi\colon \mathbb{T} \times \mathsf{M} \to \mathsf{E}$ satisfying $\xi(t, z) \in \mathsf{E}_z$ for each $(t, z) \in \mathbb{T} \times \mathsf{M}$, the following two statements are equivalent:*

(i) $\xi \in$ CF$\Gamma^{\text{hol}}(\mathbb{T}; \mathsf{E})$;

(ii) the map $\mathbb{T} \ni t \mapsto \xi_t \in \Gamma^{\text{hol}}(\mathsf{E})$ is measurable,

the following two statements are equivalent:

(iii) $\xi \in \mathrm{LI}\Gamma^{\mathrm{hol}}(\mathbb{T}; \mathsf{E})$;
(iv) the map $\mathbb{T} \ni t \mapsto \xi_t \in \Gamma^{\mathrm{hol}}(\mathsf{E})$ *is measurable and locally Bochner integrable, and the following two statements are equivalent:*
(v) $\xi \in \mathrm{LB}\Gamma^{\mathrm{hol}}(\mathbb{T}; \mathsf{E})$;
(vi) the map $\mathbb{T} \ni t \mapsto \xi_t \in \Gamma^{\mathrm{hol}}(\mathsf{E})$ *is measurable and locally essentially von Neumann bounded.*

Proof: (i) $\Longleftrightarrow$ (ii) For $z \in \mathsf{M}$ and $\alpha_z \in \mathsf{E}_z^*$, define $\mathrm{ev}_{\alpha_z} : \Gamma^{\mathrm{hol}}(\mathsf{E}) \to \mathbb{C}$ by $\mathrm{ev}_{\alpha_z}(\xi) = \langle \alpha_z; \xi(z) \rangle$. It is easy to show that ev_{α_z} is continuous and that the set of continuous functionals ev_{α_z}, $\alpha_z \in \mathsf{E}_z^*$, is point separating. Since $\Gamma^{\mathrm{hol}}(\mathsf{E})$ is a Suslin space by $\mathrm{CO}^{\mathrm{hol}}$-6, this part of the theorem follows in the same manner as the corresponding part of Theorem 6.3.

(iii) $\Longleftrightarrow$ (iv) Since $\Gamma^{\mathrm{hol}}(\mathsf{E})$ is complete and separable (by properties $\mathrm{CO}^{\mathrm{hol}}$-2 and $\mathrm{CO}^{\mathrm{hol}}$-4), the arguments from the corresponding part of Theorem 6.3 apply here.

(v) $\Longleftrightarrow$ (vi) We recall our discussion of von Neumann bounded sets in locally convex topological vector spaces preceding Lemma 3.1 above. With this in mind and using Lemma 4.3, this part of the proposition follows immediately. □

Since holomorphic vector bundles are smooth vector bundles (indeed, real analytic vector bundles), we have natural inclusions

$$\mathrm{LI}\Gamma^{\mathrm{hol}}(\mathbb{T}; \mathsf{E}) \subseteq \mathrm{CF}\Gamma^{\infty}(\mathbb{T}; \mathsf{E}), \qquad \mathrm{LB}\Gamma^{\mathrm{hol}}(\mathbb{T}; \mathsf{E}) \subseteq \mathrm{CF}\Gamma^{\infty}(\mathbb{T}; \mathsf{E}). \tag{6.4}$$

Moreover, by Proposition 4.2 we have the following.

Proposition 6.15 (Time-Varying Holomorphic Sections as Time-Varying Smooth Sections). *For an holomorphic vector bundle* $\pi : \mathsf{E} \to \mathsf{M}$ *and an interval* $\mathbb{T}$, *the inclusions* (6.4) *actually induce inclusions*

$$\mathrm{LI}\Gamma^{\mathrm{hol}}(\mathbb{T}; \mathsf{E}) \subseteq \mathrm{LI}\Gamma^{\infty}(\mathbb{T}; \mathsf{E}), \qquad \mathrm{LB}\Gamma^{\mathrm{hol}}(\mathbb{T}; \mathsf{E}) \subseteq \mathrm{LB}\Gamma^{\infty}(\mathbb{T}; \mathsf{E}).$$

Note that Theorem 6.14 applies, in particular, to vector fields and functions, giving the classes $\mathrm{CF}^{\mathrm{hol}}(\mathbb{T}; \mathsf{M})$, $\mathrm{LIC}^{\mathrm{hol}}(\mathbb{T}; \mathsf{M})$, and $\mathrm{LBC}^{\mathrm{hol}}(\mathbb{T}; \mathsf{M})$ of functions, and the classes $\mathrm{CF}\Gamma^{\mathrm{hol}}(\mathbb{T}; \mathsf{TM})$, $\mathrm{LI}\Gamma^{\mathrm{hol}}(\mathbb{T}; \mathsf{TM})$, and $\mathrm{LB}\Gamma^{\mathrm{hol}}(\mathbb{T}; \mathsf{TM})$ of vector fields. Unlike in the smooth case preceding and the real analytic case following, there is, in general, not an equivalent weak-$\mathscr{L}$ version of the preceding definitions and results. This is because our Theorem 4.5 on the equivalence of the $\mathrm{CO}^{\mathrm{hol}}$-topology and the corresponding weak-$\mathscr{L}$ topology holds only on Stein manifolds. Let us understand the consequences of this for what we are doing via an example.

Example 6.16 (Time-Varying Holomorphic Vector Fields on Compact Manifolds). Let M be a compact holomorphic manifold. By [7, Corollary IV.1.3], the only holomorphic functions on M are the locally constant functions. Therefore, since $\partial f = 0$ for every $f \in \mathrm{C}^{\mathrm{hol}}(\mathsf{M})$, a literal application of the definition shows that, were we to make weak-$\mathscr{L}$ characterisations of vector fields, i.e., give their properties by ascribing those properties to the functions obtained after Lie differentiation, we would have $\mathrm{CF}\Gamma^{\mathrm{hol}}(\mathbb{T}; \mathsf{TM})$, and, therefore, also $\mathrm{LI}\Gamma^{\mathrm{hol}}(\mathbb{T}; \mathsf{TM})$ and

$\mathrm{LB}\Gamma^{\mathrm{hol}}(\mathbb{T};\mathsf{TM})$, consisting of *all* maps $X\colon \mathbb{T}\times\mathsf{M}\to\mathsf{TM}$ satisfying $X(t,z)\in\mathsf{T}_z\mathsf{M}$ for all $z\in\mathsf{M}$. This is not a very useful class of vector fields. ∘

The following result summarises the various ways of verifying the measurability, integrability, and boundedness of holomorphic time-varying vector fields, taking into account that the preceding example necessitates that we restrict our consideration to Stein manifolds.

Theorem 6.17 (Weak Characterisations of Measurability, Integrability, and Boundedness of Holomorphic Time-Varying Vector Fields). *Let* M *be a Stein manifold, let* $\mathbb{T}\subseteq\mathbb{R}$ *be a time-domain, and let* $X\colon\mathbb{T}\times\mathsf{M}\to\mathsf{TM}$ *have the property that* X_t *is an holomorphic vector field for each* $t\in\mathbb{T}$. *Then the following statements are equivalent:*
(i) $t\mapsto X_t$ *is measurable;*
(ii) $t\mapsto\mathscr{L}_{X_t}f$ *is measurable for every* $f\in\mathrm{C}^{\mathrm{hol}}(\mathsf{M})$*;*
(iii) $t\mapsto\mathrm{ev}_z\circ X_t$ *is measurable for every* $z\in\mathsf{M}$*;*
(iv) $t\mapsto\mathrm{ev}_z\circ\mathscr{L}_{X_t}f$ *is measurable for every* $f\in\mathrm{C}^{\mathrm{hol}}(\mathsf{M})$ *and every* $z\in\mathsf{M}$,
the following two statements are equivalent:
(v) $t\mapsto X_t$ *is locally Bochner integrable;*
(vi) $t\mapsto\mathscr{L}_{X_t}f$ *is locally Bochner integrable for every* $f\in\mathrm{C}^{\mathrm{hol}}(\mathsf{M})$,
and the following two statements are equivalent:
(vii) $t\mapsto X_t$ *is locally essentially von Neumann bounded;*
(viii) $t\mapsto\mathscr{L}_{X_t}f$ *is locally essentially von Neumann bounded for every* $f\in\mathrm{C}^{\mathrm{hol}}(\mathsf{M})$.

Proof: This follows from Theorem 6.14, along with Corollary 4.6. □

Now we consider flows for the class of time-varying holomorphic vector fields defined above. Let $X\in\mathrm{LI}\Gamma^{\mathrm{hol}}(\mathbb{T};\mathsf{TM})$. According to Proposition 6.15, we can define the flow of X just as in the real case, and we shall continue to use the notation $D_X\subseteq\mathbb{T}\times\mathbb{T}\times\mathsf{M}$, Φ^X_{t,t_0}, and $\Phi^X\colon D_X\to\mathsf{M}$ as in the smooth case. The following result provides the attributes of the flow in the holomorphic case. This result follows easily from the constructions in the usual existence and uniqueness theorem for ordinary differential equations, but we could not find the result explicitly in the literature for measurable time dependence. Thus we provide the details here.

Theorem 6.18 (Flows of Vector Fields from $\mathrm{LI}\Gamma^{\mathrm{hol}}(\mathbb{T};\mathsf{TM})$). *Let* M *be an holomorphic manifold, let* $\mathbb{T}$ *be an interval, and let* $X\in\mathrm{LI}\Gamma^{\mathrm{hol}}(\mathbb{T};\mathsf{TM})$. *Then there exist a subset* $D_X\subseteq\mathbb{T}\times\mathbb{T}\times\mathsf{M}$ *and a map* $\Phi^X\colon D_X\to\mathsf{M}$ *with the following properties for each* $(t_0,z_0)\in\mathbb{T}\times\mathsf{M}$*:*
(i) the set

$$\mathbb{T}_X(t_0,z_0)=\{t\in\mathbb{T}\mid (t,t_0,z_0)\in D_X\}$$

is an interval;
(ii) there exists a locally absolutely continuous curve $t\mapsto\xi(t)$ *satisfying*

$$\xi'(t)=X(t,\xi(t)),\quad \xi(t_0)=z_0,$$

for almost all $t\in|t_0,t_1|$ *if and only if* $t_1\in\mathbb{T}_X(t_0,z_0)$*;*

(iii) $\frac{\mathrm{d}}{\mathrm{d}t}\Phi^X(t, t_0, z_0) = X(t, \Phi^X(t, t_0, z_0))$ *for almost all* $t \in \mathbb{T}_X(t_0, z_0)$*;*
(iv) for each $t \in \mathbb{T}$ *for which* $(t, t_0, z_0) \in D_X$*, there exists a neighbourhood* $\mathcal{U}$ *of* z_0 *such that the mapping* $z \mapsto \Phi^X(t, t_0, z)$ *is defined and of class* C^{hol} *on* $\mathcal{U}$*.*

Proof: Given Proposition 6.15, the only part of the theorem that does not follow from Theorem 6.6 is the holomorphic dependence on initial conditions. This is a local assertion, so we let $(\mathcal{U}, \phi)$ be an holomorphic chart for M with coordinates denoted by $(z^1, \dots, z^n)$. We denote by $\boldsymbol{X}\colon \mathbb{T} \times \phi(\mathcal{U}) \to \mathbb{C}^n$ the local representative of X. By Proposition 6.15, this local representative is locally integrally C^∞-bounded. To prove holomorphicity of the flow, we recall the construction for the existence and uniqueness theorem for the solutions of the initial value problem

$$\dot{\gamma}(t) = \boldsymbol{X}(t, \gamma(t)), \qquad \gamma(t_0) = z,$$

see, e.g., [15, Sect. 1.2]. On some suitable product domain $\mathbb{T}' \times \mathsf{B}(r, z_0)$ (the ball being contained in $\phi(\mathcal{U}) \subseteq \mathbb{C}^n$) we denote by $C^0(\mathbb{T}' \times \mathsf{B}(r, z_0); \mathbb{C}^n)$ the Banach space of continuous mappings with the ∞-norm [9, Theorem 7.9]. We define an operator

$$\Phi\colon C^0(\mathbb{T}' \times \mathsf{B}(r, z_0); \mathbb{C}^n) \to C^0(\mathbb{T}' \times \mathsf{B}(r, z_0); \mathbb{C}^n)$$

by

$$\Phi(\gamma)(t, z) = z + \int_{t_0}^{t} \boldsymbol{X}(s, \gamma(s, z))\,\mathrm{d}s.$$

One shows that this mapping, with domains suitably defined, is a contraction mapping, and so, by iterating the mapping, one constructs a sequence in $C^0(\mathbb{T}' \times \mathsf{B}(r, z_0); \mathbb{C}^n)$ converging to a fixed point, and the fixed point, necessarily satisfying

$$\gamma(t, z) = z + \int_{t_0}^{t} \boldsymbol{X}(s, \gamma(s, z))\,\mathrm{d}s$$

and $\gamma(t_0, z) = z$, has the property that $\gamma(t, z) = \Phi^X(t, t_0, z)$.

Let us consider the sequence one constructs in this procedure. We define $\gamma_0 \in C^0(\mathbb{T}' \times \mathsf{B}(r, z_0); \mathbb{C}^n)$ by $\gamma_0(t, z) = z$. Certainly γ_0 is holomorphic in z. Now define $\gamma_1 \in C^0(\mathbb{T}' \times \mathsf{B}(r, z_0); \mathbb{C}^n)$ by

$$\gamma_1(t, z) = \Phi(\gamma_0) = z + \int_{t_0}^{t} \boldsymbol{X}(s, z)\,\mathrm{d}s.$$

Since $\boldsymbol{X} \in \mathrm{LI}\Gamma^{\mathrm{hol}}(\mathbb{T}'; \mathsf{TB}(r, z_0))$, we have

$$\frac{\partial}{\partial \bar{z}^j}\gamma_1(t, z) = \frac{\partial}{\partial \bar{z}^j} z + \int_{t_0}^{t} \frac{\partial}{\partial \bar{z}^j}\boldsymbol{X}(s, \gamma_0(s, z))\,\mathrm{d}s = 0, \qquad j \in \{1, \dots, n\},$$

swapping the derivative and the integral by the Dominated Convergence Theorem [10, Theorem 16.11] (also noting by Proposition 6.15 that derivatives of $\boldsymbol{X}$ are bounded by an integrable function). Thus γ_1 is holomorphic for each fixed $t \in \mathbb{T}'$. By iterating with t fixed, we have a sequence $(\gamma_{j,t})_{j \in \mathbb{Z}_{\geq 0}}$ of holomorphic mappings from $\mathsf{B}(r, z_0)$ converging uniformly to the function $\boldsymbol{\gamma}$ that describes how the solution at time t depends on the initial condition z. The limit function is necessarily holomorphic [8, p. 5]. □

6.4 The Real Analytic Case

Let us now turn to describing real analytic time-varying sections. We thus will consider a real analytic vector bundle $\pi\colon \mathsf{E} \to \mathsf{M}$ with ∇^0 a real analytic linear connection on E, ∇ a real analytic affine connection on M, $\mathbb{G}_0$ a real analytic fibre metric on E, and $\mathbb{G}$ a real analytic Riemannian metric on M. This defines the seminorms $p^{\omega}_{K,\boldsymbol{a}}$, $K \subseteq \mathsf{M}$ compact, $\boldsymbol{a} \in c_0(\mathbb{Z}_{\geq 0}; \mathbb{R}_{>0})$, describing the C^{ω}-topology as in Theorem 5.5.

Definition 6.19 (Real Analytic Carathéodory Section). Let $\pi\colon \mathsf{E} \to \mathsf{M}$ be a real analytic vector bundle and let $\mathbb{T} \subseteq \mathbb{R}$ be an interval. A ***Carathéodory section of class C^{ω}*** of E is a map $\xi\colon \mathbb{T} \times \mathsf{M} \to \mathsf{E}$ with the following properties:
(i) $\xi(t, x) \in \mathsf{E}_x$ for each $(t, x) \in \mathbb{T} \times \mathsf{M}$;
(ii) for each $t \in \mathbb{T}$, the map $\xi_t\colon \mathsf{M} \to \mathsf{E}$ defined by $\xi_t(x) = \xi(t, x)$ is of class C^{ω};
(iii) for each $x \in \mathsf{M}$, the map $\xi^x\colon \mathbb{T} \to \mathsf{E}$ defined by $\xi^x(t) = \xi(t, x)$ is Lebesgue measurable.

We shall call $\mathbb{T}$ the ***time-domain*** for the section. By $\mathrm{CF}\Gamma^{\omega}(\mathbb{T}; \mathsf{E})$ we denote the set of Carathéodory sections of class C^{ω} of E. ○

Now we turn to placing restrictions on the time dependence to allow us to do useful things.

Definition 6.20 (Locally Integrally C^{ω}-Bounded and Locally Essentially C^{ω}-Bounded Sections). Let $\pi\colon \mathsf{E} \to \mathsf{M}$ be a real analytic vector bundle and let $\mathbb{T} \subseteq \mathbb{R}$ be an interval. A Carathéodory section $\xi\colon \mathbb{T} \times \mathsf{M} \to \mathsf{E}$ of class C^{ω} is
(i) ***locally integrally C^{ω}-bounded*** if, for every compact set $K \subseteq \mathsf{M}$ and every $\boldsymbol{a} \in c_0(\mathbb{Z}_{\geq 0}; \mathbb{R}_{>0})$, there exists $g \in \mathrm{L}^1_{\mathrm{loc}}(\mathbb{T}; \mathbb{R}_{\geq 0})$ such that

$$a_0 a_1 \cdots a_m \|j_m \xi_t(x)\|_{\overline{\mathbb{G}}_m} \leq g(t), \qquad (t, x) \in \mathbb{T} \times K,\ m \in \mathbb{Z}_{\geq 0},$$

and is
(ii) ***locally essentially C^{ω}-bounded*** if, for every compact set $K \subseteq \mathsf{M}$ and every $\boldsymbol{a} \in c_0(\mathbb{Z}_{\geq 0}; \mathbb{R}_{>0})$, there exists $g \in \mathrm{L}^{\infty}_{\mathrm{loc}}(\mathbb{T}; \mathbb{R}_{\geq 0})$ such that

$$a_0 a_1 \cdots a_m \|j_m \xi_t(x)\|_{\overline{\mathbb{G}}_m} \leq g(t), \qquad (t, x) \in \mathbb{T} \times K,\ m \in \mathbb{Z}_{\geq 0}.$$

The set of locally integrally C^{ω}-bounded sections of E with time-domain $\mathbb{T}$ is denoted by $\mathrm{LI}\Gamma^{\omega}(\mathbb{T}, \mathsf{E})$ and the set of locally essentially C^{ω}-bounded sections of E with time-domain $\mathbb{T}$ is denoted by $\mathrm{LB}\Gamma^{\omega}(\mathbb{T}; \mathsf{E})$. ○

As with smooth and holomorphic sections, the preceding definitions admit topological characterisations.

Theorem 6.21 (Topological Characterisation of Real Analytic Carathéodory Sections). *Let $\pi\colon \mathsf{E} \to \mathsf{M}$ be a real analytic manifold and let $\mathbb{T} \subseteq \mathbb{R}$ be an interval. For a map $\xi\colon \mathbb{T}\times\mathsf{M} \to \mathsf{E}$ satisfying $\xi(t,x) \in \mathsf{E}_x$ for each $(t,x) \in \mathbb{T}\times\mathsf{M}$, the following two statements are equivalent:*
(i) $\xi \in \mathrm{CF}\Gamma^{\omega}(\mathbb{T};\mathsf{E})$;
(ii) the map $\mathbb{T} \ni t \mapsto \xi_t \in \Gamma^{\omega}(\mathsf{E})$ is measurable,
the following two statements are equivalent:
(iii) $\xi \in \mathrm{LI}\Gamma^{\omega}(\mathbb{T};\mathsf{E})$;
(iv) the map $\mathbb{T} \ni t \mapsto \xi_t \in \Gamma^{\omega}(\mathsf{E})$ is measurable and locally Bochner integrable,
and the following two statements are equivalent:
(v) $\xi \in \mathrm{LB}\Gamma^{\omega}(\mathbb{T};\mathsf{E})$;
(vi) the map $\mathbb{T} \ni t \mapsto \xi_t \in \Gamma^{\omega}(\mathsf{E})$ is measurable and locally essentially von Neumann bounded.

Proof: Just as in the smooth case in Theorem 6.3, this is deduced from the following facts: (1) evaluation maps ev_{α_x}, $\alpha_x \in \mathsf{E}^*$, are continuous and point separating; (2) $\Gamma^{\omega}(\mathsf{E})$ is a Suslin space (property C^{ω}-6); (3) $\Gamma^{\omega}(\mathsf{E})$ is complete and separable (properties C^{ω}-2 and C^{ω}-4; (4) we understand von Neumann bounded subsets of $\Gamma^{\omega}(\mathsf{E})$ by Lemma 5.6. □

Note that Theorem 6.21 applies, in particular, to vector fields and functions, giving the classes $\mathrm{CF}^{\omega}(\mathbb{T};\mathsf{M})$, $\mathrm{LIC}^{\omega}(\mathbb{T};\mathsf{M})$, and $\mathrm{LBC}^{\omega}(\mathbb{T};\mathsf{M})$ of functions, and the classes $\mathrm{CF}\Gamma^{\omega}(\mathbb{T};\mathsf{TM})$, $\mathrm{LI}\Gamma^{\omega}(\mathbb{T};\mathsf{TM})$, and $\mathrm{LB}\Gamma^{\omega}(\mathbb{T};\mathsf{TM})$ of vector fields. The following result then summarises the various ways of verifying the measurability, integrability, and boundedness of real analytic time-varying vector fields.

Theorem 6.22 (Weak Characterisations of Measurability, Integrability, and Boundedness of Real Analytic Time-Varying Vector Fields). *Let M be a real analytic manifold, let $\mathbb{T} \subseteq \mathbb{R}$ be a time-domain, and let $X\colon \mathbb{T}\times\mathsf{M} \to \mathsf{TM}$ have the property that X_t is a real analytic vector field for each $t \in \mathbb{T}$. Then the following statements are equivalent:*
(i) $t \mapsto X_t$ is measurable;
(ii) $t \mapsto \mathscr{L}_{X_t} f$ is measurable for every $f \in \mathrm{C}^{\omega}(\mathsf{M})$;
(iii) $t \mapsto \mathrm{ev}_x \circ X_t$ is measurable for every $x \in \mathsf{M}$;
(iv) $t \mapsto \mathrm{ev}_x \circ \mathscr{L}_{X_t} f$ is measurable for every $f \in \mathrm{C}^{\omega}(\mathsf{M})$ and every $x \in \mathsf{M}$,
the following two statements are equivalent:
(v) $t \mapsto X_t$ is locally Bochner integrable;
(vi) $t \mapsto \mathscr{L}_{X_t} f$ is locally Bochner integrable for every $f \in \mathrm{C}^{\omega}(\mathsf{M})$,
and the following two statements are equivalent:
(vii) $t \mapsto X_t$ is locally essentially bounded;
(viii) $t \mapsto \mathscr{L}_{X_t} f$ is locally essentially bounded in the von Neumann bornology for every $f \in \mathrm{C}^{\omega}(\mathsf{M})$.

Proof: This follows from Theorem 6.21, along with Corollary 5.9. □

Let us verify that real analytic time-varying sections have the expected relationship to their smooth brethren.

Proposition 6.23 (Time-Varying Real Analytic Sections as Time-Varying Smooth Sections). *For a real analytic vector bundle $\pi\colon \mathsf{E} \to \mathsf{M}$ and an interval $\mathbb{T}$, we have*

$$\mathrm{LI}\Gamma^{\omega}(\mathbb{T};\mathsf{E}) \subseteq \mathrm{LI}\Gamma^{\infty}(\mathbb{T};\mathsf{E}), \qquad \mathrm{LB}\Gamma^{\omega}(\mathbb{T};\mathsf{M}) \subseteq \mathrm{LB}\Gamma^{\infty}(\mathbb{T};\mathsf{M}).$$

Proof: It is obvious that real analytic Carathéodory sections are smooth Carathéodory sections.

Let us verify only that $\mathrm{LI}\Gamma^{\omega}(\mathbb{T};\mathsf{E}) \subseteq \mathrm{LI}\Gamma^{\infty}(\mathbb{T};\mathsf{E})$, as the essentially bounded case follows in the same manner. We let $K \subseteq \mathsf{M}$ be compact and let $m \in \mathbb{Z}_{\geq 0}$. Choose (arbitrarily) $\boldsymbol{a} \in c_0(\mathbb{Z}_{\geq 0};\mathbb{R}_{>0})$. Then, if $\xi \in \mathrm{LI}\Gamma^{\omega}(\mathbb{T};\mathsf{E})$, there exists $g \in \mathrm{L}^1_{\mathrm{loc}}(\mathbb{T};\mathbb{R}_{\geq 0})$ such that

$$a_0 a_1 \cdots a_m \|j_m\xi_t(x)\|_{\overline{\mathbb{G}}_m} \leq g(t), \qquad x \in K,\ t \in \mathbb{T},\ m \in \mathbb{Z}_{\geq 0}.$$

Thus, taking $g_{\boldsymbol{a},m} \in \mathrm{L}^1_{\mathrm{loc}}(\mathbb{T};\mathbb{R}_{\geq 0})$ defined by

$$g_{\boldsymbol{a},m}(t) = \frac{1}{a_0 a_1 \cdots a_m} g(t),$$

we have

$$\|j_m\xi_t(x)\|_{\overline{\mathbb{G}}_m} \leq g_{\boldsymbol{a},m}(t), \qquad x \in K,\ t \in \mathbb{T}$$

showing that $\xi \in \mathrm{LI}\Gamma^{\infty}(\mathbb{T};\mathsf{E})$. □

Having understood the comparatively simple relationship between real analytic and smooth time-varying sections, let us consider the correspondence between real analytic and holomorphic time-varying sections. First, note that if $\mathbb{T} \subseteq \mathbb{R}$ is an interval and if $\overline{\mathcal{U}} \in \mathscr{N}_{\mathsf{M}}$ is a neighbourhood of M in a complexification $\overline{\mathsf{M}}$, then we have an inclusion

$$\begin{aligned}\rho_{\overline{\mathcal{U}},\mathsf{M}}\colon \mathrm{CF}\Gamma^{\mathrm{hol},\mathbb{R}}(\mathbb{T};\overline{\mathsf{E}}|\overline{\mathcal{U}}) &\to \mathrm{CF}\Gamma^{\omega}(\mathbb{T};\mathsf{E})\\ \overline{\xi} &\mapsto \overline{\xi}|\mathsf{M}.\end{aligned}$$

(Here the notation $\mathrm{CF}\Gamma^{\mathrm{hol},\mathbb{R}}(\mathbb{T};\overline{\mathsf{E}}|\overline{\mathcal{U}})$ refers to those Carathéodory sections that are real when restricted to M, cf. the constructions of Sect. 5.1.2.) However, this inclusion does not characterise all real analytic Carathéodory sections, as the following example shows.

Example 6.24 (A Real Analytic Carathéodory Function Not Extending to One That Is Holomorphic). Let $\mathbb{T}$ be any interval for which $0 \in \mathrm{int}(\mathbb{T})$. We consider the real analytic Carathédory function on $\mathbb{R}$ with time-domain $\mathbb{T}$ defined by

$$f(t,x) = \begin{cases} \frac{t^2}{t^2+x^2}, & t \neq 0,\\ 0, & t = 0.\end{cases}$$

It is clear that $x \mapsto f(t,x)$ is real analytic for every $t \in \mathbb{T}$ and that $t \mapsto f(t,x)$ is measurable for every $x \in \mathbb{R}$. We claim that there is no neighbourhood $\overline{\mathcal{U}} \subseteq \mathbb{C}$ of

$\overline{\mathbb{R}} \subseteq \mathbb{C}$ such that f is the restriction to $\mathbb{R}$ of an holomorphic Carathéodory function on $\overline{\mathcal{U}}$. Indeed, let $\overline{\mathcal{U}} \subseteq \mathbb{C}$ be a neighbourhood of $\mathbb{R}$ and choose $t \in \mathbb{R}_{>0}$ sufficiently small that $\overline{\mathsf{D}}(t,0) \subseteq \overline{\mathcal{U}}$. Note that $f_t\colon x \mapsto \frac{1}{1+(x/t)^2}$ does not admit an holomorphic extension to any open set containing $\overline{\mathsf{D}}(t,0)$ since the radius of convergence of $z \mapsto \frac{1}{1+(z/t)^2}$ is t, cf. the discussion at the beginning of Chap. 5. Note that our construction actually shows that in no neighbourhood of $(0,0) \in \mathbb{R}\times\mathbb{R}$ is there an holomorphic extension of f. ○

Fortunately, the example will not bother us, although it does serve to illustrate that the following result is not immediate.

Theorem 6.25 (Real Analytic Time-Varying Vector Fields as Restrictions of Holomorphic Time-Varying Vector Fields). *Let $\pi\colon \mathsf{E} \to \mathsf{M}$ be a real analytic vector bundle with complexification $\overline{\pi}\colon \overline{\mathsf{E}} \to \overline{\mathsf{M}}$, and let $\mathbb{T}$ be a time-domain. For a map $\xi\colon \mathbb{T}\times\mathsf{M} \to \mathsf{E}$ satisfying $\xi(t,x) \in \mathsf{E}_x$ for every $(t,x) \in \mathbb{T}\times\mathsf{M}$, the following statements hold:*

(i) if $\xi \in \mathrm{LI}\Gamma^{\omega}(\mathbb{T};\mathsf{E})$, then, for each $(t_0,x_0) \in \mathbb{T}\times\mathsf{M}$ and each bounded subinterval $\mathbb{T}' \subseteq \mathbb{T}$ containing t_0, there exist a neighbourhood $\overline{\mathcal{U}}$ of x_0 in $\overline{\mathsf{M}}$ and $\overline{\xi} \in \Gamma^{\mathrm{hol}}(\mathbb{T}';\overline{\mathsf{E}}|\overline{\mathcal{U}})$ such that $\overline{\xi}(t,x) = \xi(t,x)$ for each $t \in \mathbb{T}'$ and $x \in \overline{\mathcal{U}}\cap\mathsf{M}$;

(ii) if, for each $x_0 \in \mathsf{M}$, there exist a neighbourhood $\overline{\mathcal{U}}$ of x_0 in $\overline{\mathsf{M}}$ and $\overline{\xi} \in \Gamma^{\mathrm{hol}}(\mathbb{T};\overline{\mathsf{E}}|\overline{\mathcal{U}})$ such that $\overline{\xi}(t,x) = \xi(t,x)$ for each $t \in \mathbb{T}$ and $x \in \overline{\mathcal{U}}\cap\mathsf{M}$, then $\xi \in \mathrm{LI}\Gamma^{\omega}(\mathbb{T};\mathsf{E})$.

Proof: (i) We let $\mathbb{T}' \subseteq \mathbb{T}$ be a bounded subinterval containing t_0 and let $\mathcal{U}$ be a relatively compact neighbourhood of x_0. Let $(\overline{\mathcal{U}}_j)_{j\in\mathbb{Z}_{>0}}$ be a sequence of neighbourhoods of $\mathrm{cl}(\mathcal{U})$ in $\overline{\mathsf{M}}$ with the properties that $\mathrm{cl}(\overline{\mathcal{U}}_{j+1}) \subseteq \overline{\mathcal{U}}_j$ and that $\cap_{j\in\mathbb{Z}_{>0}}\overline{\mathcal{U}}_j = \mathrm{cl}(\mathcal{U})$. We first note that

$$\mathrm{L}^1(\mathbb{T}';\Gamma^{\mathrm{hol},\mathbb{R}}(\overline{\mathsf{E}}|\overline{\mathcal{U}}_j)) \simeq \mathrm{L}^1(\mathbb{T}';\mathbb{R})\widehat{\otimes}_{\pi}\Gamma^{\mathrm{hol},\mathbb{R}}(\overline{\mathsf{E}}|\overline{\mathcal{U}}_j),$$

with $\widehat{\otimes}_{\pi}$ denoting the completed projective tensor product [14, Theorem III.6.5]. The theorem of Schaefer and Wolff [14] is given for Banach spaces, and they also assert the validity of this for locally convex spaces; thus we also have

$$\mathrm{L}^1(\mathbb{T}';\mathscr{G}^{\mathrm{hol},\mathbb{R}}_{\mathrm{cl}(\mathcal{U}),\overline{\mathsf{E}}}) \simeq \mathrm{L}^1(\mathbb{T}';\mathbb{R})\widehat{\otimes}_{\pi}\mathscr{G}^{\mathrm{hol},\mathbb{R}}_{\mathrm{cl}(\mathcal{U}),\overline{\mathsf{E}}}.$$

In both cases, the isomorphisms are in the category of locally convex topological vector spaces. We now claim

$$\mathrm{L}^1(\mathbb{T}';\mathbb{R})\otimes_{\pi}\mathscr{G}^{\mathrm{hol},\mathbb{R}}_{\mathrm{cl}(\mathcal{U}),\overline{\mathsf{E}}}$$

is the direct limit of the directed system

$$(\mathrm{L}^1(\mathbb{T}';\mathbb{R})\otimes_{\pi}\Gamma^{\mathrm{hol},\mathbb{R}}(\overline{\mathsf{E}}|\overline{\mathcal{U}}_j))_{j\in\mathbb{Z}_{>0}}$$

with the associated mappings $\mathrm{id}\otimes_{\pi} r_{\mathrm{cl}(\mathcal{U}),j}$, $j \in \mathbb{Z}_{>0}$, where $r_{\mathrm{cl}(\mathcal{U}),j}$ is defined as in (5.1). (Here $\otimes_{\pi}$ is the uncompleted projective tensor product.) We, moreover, claim that the direct limit topology is boundedly retractive, meaning that bounded sets in the direct limit are contained in and bounded in a single component of the directed system and, moreover, the topology on the bounded set induced by the component is the same as that induced by the direct limit.

Results of this sort have been the subject of research in the area of locally convex topologies, with the aim being to deduce conditions on the structure of the spaces comprising the directed system, and on the corresponding mappings (for us, the inclusion mappings and their tensor products with the identity on $L^1(\mathbb{T}';\mathbb{R})$), that ensure that direct limits commute with tensor product, and that the associated direct limit topology is boundedly retractive. We shall make principal use of the results given in [11]. To state the arguments with at least a little context, let us reproduce two conditions used in [11].

Condition (M) of Retakh [12] Let $(\mathsf{V}_j)_{j\in\mathbb{Z}_{>0}}$ be a directed system of locally convex spaces with strict direct limit V. The direct limit topology of V satisfies ***condition (M)*** if there exists a sequence $(\mathcal{O}_j)_{j\in\mathbb{Z}_{>0}}$ for which

(i) $\mathcal{O}_j$ is a balanced convex neighbourhood of $0 \in \mathsf{V}_j$,
(ii) $\mathcal{O}_j \subseteq \mathcal{O}_{j+1}$ for each $j \in \mathbb{Z}_{>0}$, and
(iii) for every $j \in \mathbb{Z}_{>0}$, there exists $k \geq j$ such that the topology induced on $\mathcal{O}_j$ by its inclusion in V_k and its inclusion in V agree. ∘

Condition (MO) of Mangino [11] Let $(\mathsf{V}_j)_{j\in\mathbb{Z}_{>0}}$ be a directed system of metrisable locally convex spaces with strict direct limit V. Let $i_{j,k}\colon \mathsf{V}_j \to \mathsf{V}_k$ be the inclusion for $k \geq j$ and let $i_j\colon \mathsf{V}_j \to \mathsf{V}$ be the induced map into the direct limit.

Suppose that, for each $j \in \mathbb{Z}_{>0}$, we have a sequence $(p_{j,l})_{l\in\mathbb{Z}_{>0}}$ of seminorms defining the topology of V_j such that $p_{j,l_1} \geq p_{j,l_2}$ if $l_1 \geq l_2$. Let

$$\mathsf{V}_{j,l} = \mathsf{V}_j/\{v \in \mathsf{V}_j \mid p_{j,l}(v) = 0\}$$

and denote by $\hat{p}_{j,l}$ the norm on $\mathsf{V}_{j,l}$ induced by $p_{j,l}$ [14, p. 97]. Let $\pi_{j,l}\colon \mathsf{V}_j \to \mathsf{V}_{j,l}$ be the canonical projection. Let $\overline{\mathsf{V}}_{j,l}$ be the completion of $\mathsf{V}_{j,l}$. The family $(\overline{\mathsf{V}}_{j,l})_{j,l\in\mathbb{Z}_{>0}}$ is called a ***projective spectrum*** for V_j. Denote

$$\mathcal{O}_{j,l} = \{v \in \mathsf{V}_j \mid p_{j,l}(v) \leq 1\}.$$

The direct limit topology of V satisfies ***condition (MO)*** if there exists a sequence $(\mathcal{O}_j)_{j\in\mathbb{Z}_{>0}}$ and if, for every $j \in \mathbb{Z}_{>0}$, there exists a projective spectrum $(\overline{\mathsf{V}}_{j,l})_{j,l\in\mathbb{Z}_{>0}}$ for V_j for which

(i) $\mathcal{O}_j$ is a balanced convex neighbourhood of $0 \in \mathsf{V}_j$,
(ii) $\mathcal{O}_j \subseteq \mathcal{O}_{j+1}$ for each $j \in \mathbb{Z}_{>0}$, and
(iii) for every $j \in \mathbb{Z}_{>0}$, there exists $k \geq j$ such that, for every $l \in \mathbb{Z}_{>0}$, there exists $A \in L(\mathsf{V}; \overline{\mathsf{V}}_{k,l})$ satisfying

$$(\pi_{k,l} \circ i_{j,k} - A \circ i_j)(\mathcal{O}_j) \subseteq \mathrm{cl}(\pi_{k,l}(\mathsf{O}_{k,l})),$$

the closure on the right being taken in the norm topology of $\overline{\mathsf{V}}_{k,l}$. ∘

With these concepts, we have the following statements. We let $(\mathsf{V}_j)_{j\in\mathbb{Z}_{>0}}$ be a directed system of metrisable locally convex spaces with strict direct limit V.

1. If the direct limit topology on V satisfies condition (MO), then, for any Banach space U, $\mathsf{U}\otimes_\pi\mathsf{V}$ is the direct limit of the directed system $(\mathsf{U}\otimes_\pi\mathsf{V}_j)_{j\in\mathbb{Z}_{>0}}$, and the direct limit topology on $\mathsf{U}\otimes_\pi\mathsf{V}$ satisfies condition (M) [11, Theorem 1.3].
2. If the spaces V_j, $j\in\mathbb{Z}_{>0}$, are nuclear and if the direct limit topology on V is regular, then the direct limit topology on V satisfies condition (MO) [11, Theorem 1.3].
3. If the direct limit topology on V satisfies condition (M), then this direct limit topology is boundedly retractive [19].

Using these arguments we make the following conclusions.

4. The direct limit topology on $\mathscr{G}^{\mathrm{hol},\mathbb{R}}_{\mathrm{cl}(\mathcal{U}),\overline{\mathsf{E}}}$ satisfies condition (MO) (by virtue of assertion 2 above and by the properties of the direct limit topology enunciated in Sect. 5.3, specifically that the direct limit is a regular direct limit of nuclear Fréchet spaces).
5. The space $\mathrm{L}^1(\mathbb{T}';\mathbb{R})\otimes_\pi\mathscr{G}^{\mathrm{hol},\mathbb{R}}_{\mathrm{cl}(\mathcal{U}),\overline{\mathsf{E}}}$ is the direct limit of the directed sequence $(\mathrm{L}^1(\mathbb{T}';\mathbb{R})\otimes_\pi\Gamma^{\mathrm{hol},\mathbb{R}}(\overline{\mathsf{E}}|\overline{\mathcal{U}}_{\mathrm{j}}))_{\mathrm{j}\in\mathbb{Z}_{>0}}$ (by virtue of assertion 1 above).
6. The direct limit topology on $\mathrm{L}^1(\mathbb{T}';\mathbb{R})\otimes_\pi\mathscr{G}^{\mathrm{hol},\mathbb{R}}_{\mathrm{cl}(\mathcal{U}),\overline{\mathsf{E}}}$ satisfies condition (M) (by virtue of assertion 1 above).
7. The direct limit topology on $\mathrm{L}^1(\mathbb{T}';\mathbb{R})\otimes_\pi\mathscr{G}^{\mathrm{hol},\mathbb{R}}_{\mathrm{cl}(\mathcal{U}),\overline{\mathsf{E}}}$ is boundedly retractive (by virtue of assertion 3 above).

We shall also need the following lemma.

Lemma 1 *Let $K\subseteq\mathsf{M}$ be compact. If $[\overline{\xi}]_K\in\mathrm{L}^1(\mathbb{T}';\mathscr{G}^{\mathrm{hol},\mathbb{R}}_{\mathsf{K},\overline{\mathsf{E}}})$, then there exists a sequence $([\overline{\xi}_k]_K)_{k\in\mathbb{Z}_{>0}}$ in $\mathrm{L}^1(\mathbb{T}';\mathbb{R})\otimes\mathscr{G}^{\mathrm{hol},\mathbb{R}}_{\mathsf{K},\overline{\mathsf{E}}}$ converging to $[\overline{\xi}]_K$ in the topology of $\mathrm{L}^1(\mathbb{T}';\mathscr{G}^{\mathrm{hol}}_{\mathsf{K},\overline{\mathsf{E}}})$.*

Proof: Since $\mathrm{L}^1(\mathbb{T}';\mathscr{G}^{\mathrm{hol},\mathbb{R}}_{\mathsf{K},\overline{\mathsf{E}}})$ is the completion of $\mathrm{L}^1(\mathbb{T}';\mathbb{R})\otimes_\pi\mathscr{G}^{\mathrm{hol},\mathbb{R}}_{\mathsf{K},\overline{\mathsf{E}}}$, there exists a *net* $([\overline{\xi}_i]_K)_{i\in I}$ converging to $[\overline{\xi}]$, so the conclusion here is that we can actually find a converging *sequence*.

To prove this we argue as follows. Recall properties $\mathscr{G}^{\mathrm{hol},\mathbb{R}}$-5 and $\mathscr{G}^{\mathrm{hol},\mathbb{R}}$-6 of $\mathscr{G}^{\mathrm{hol},\mathbb{R}}_{K,\overline{\mathsf{E}}}$, indicating that it is reflexive and its dual is a nuclear Fréchet space. Thus $\mathscr{G}^{\mathrm{hol},\mathbb{R}}_{K,\overline{\mathsf{E}}}$ is the dual of a nuclear Fréchet space. Also recall from property $\mathscr{G}^{\mathrm{hol},\mathbb{R}}$-8 that $\mathscr{G}^{\mathrm{hol},\mathbb{R}}_{K,\overline{\mathsf{E}}}$ is a Suslin space. Now, by combining [18, Theorem 7] with remark (1) at the bottom of page 76 of [18] (and being aware that Bochner integrability as defined in [18] is not a priori the same as Bochner integrability as we mean it), there exists a sequence $([\overline{\xi}_k]_K)_{k\in\mathbb{Z}_{>0}}$ of simple functions, i.e., elements of $\mathrm{L}^1(\mathbb{T}';\mathbb{R})\otimes\mathscr{G}^{\mathrm{hol},\mathbb{R}}_{\mathsf{K},\overline{\mathsf{E}}}$, such that

$$\lim_{k\to\infty}[\overline{\xi}_k(t)]_K=[\overline{\xi}(t)]_K,\qquad \text{a.e. } t\in\mathbb{T}',$$

(this limit being in the topology of $\mathscr{G}^{\mathrm{hol},\mathbb{R}}_{K,\overline{\mathsf{E}}}$) and

$$\lim_{k\to\infty}\int_{\mathbb{T}'}([\overline{\xi}(t)]_K - [\overline{\xi}_k(t)]_K)\,\mathrm{d}t = 0.$$

This implies, by the Dominated Convergence Theorem, that

$$\lim_{k\to\infty}\int_{\mathbb{T}'} p^{\omega}_{K,\boldsymbol{a}}([\overline{\xi}(t)]_K - [\overline{\xi}_k(t)]_K)\,\mathrm{d}t = 0$$

for every $\boldsymbol{a}\in c_0(\mathbb{Z}_{\geq 0};\mathbb{R}_{>0})$, giving convergence in

$$\mathrm{L}^1(\mathbb{T}';\mathscr{G}^{\mathrm{hol},\mathbb{R}}_{\mathrm{K},\overline{\mathsf{E}}}) \simeq \mathrm{L}^1(\mathbb{T}';\mathbb{R})\widehat{\otimes}_\pi\mathscr{G}^{\mathrm{hol},\mathbb{R}}_{\mathrm{K},\overline{\mathsf{E}}},$$

as desired. ▽

The remainder of the proof is straightforward. Since $\xi\in \mathrm{LI}\Gamma^{\omega}(\mathbb{T};\mathsf{E})$, the map

$$\mathbb{T}'\ni t\mapsto \xi_t\in\Gamma^{\omega}(\mathsf{E})$$

is an element of $\mathrm{L}^1(\mathbb{T}';\Gamma^{\omega}(\mathsf{E}))$ by Theorem 6.21. Therefore, if $[\overline{\xi}]_{\mathrm{cl}(\mathcal{U})}$ is the image of ξ under the natural mapping from $\Gamma^{\omega}(\mathsf{E})$ to $\mathscr{G}^{\mathrm{hol},\mathbb{R}}_{\mathrm{cl}(\mathcal{U}),\overline{\mathsf{E}}}$, the map

$$\mathbb{T}'\ni t\mapsto [\overline{\xi(t)}]_{\mathrm{cl}(\mathcal{U})}\in\mathscr{G}^{\mathrm{hol},\mathbb{R}}_{\mathrm{cl}(\mathcal{U}),\overline{\mathsf{E}}}$$

is an element of $\mathrm{L}^1(\mathbb{T}';\mathscr{G}^{\mathrm{hol},\mathbb{R}}_{\mathrm{cl}(\mathcal{U}),\overline{\mathsf{E}}})$, since continuous linear maps commute with integration [4, Lemma 1.2]. Therefore, by the Lemma above, there exists a sequence $([\overline{\xi}_k]_{\mathrm{cl}(\mathcal{U})})_{k\in\mathbb{Z}_{>0}}$ in $\mathrm{L}^1(\mathbb{T}';\mathbb{R})\otimes\mathscr{G}^{\mathrm{hol},\mathbb{R}}_{\mathrm{cl}(\mathcal{U}),\overline{\mathsf{E}}}$ that converges to $[\overline{\xi}]_{\mathrm{cl}(\mathcal{U})}$. By our conclusion 5 above, the topology in which this convergence takes place is the completion of the direct limit topology associated with the directed system $(\mathrm{L}^1(\mathbb{T}';\mathbb{R})\otimes_\pi \Gamma^{\mathrm{hol},\mathbb{R}}(\overline{\mathsf{E}}|\overline{\mathcal{U}}_j))_{j\in\mathbb{Z}_{>0}}$. The direct limit topology on $\mathrm{L}^1(\mathbb{T}';\mathbb{R})\otimes_\pi\mathscr{G}^{\mathrm{hol},\mathbb{R}}_{\mathrm{cl}(\mathcal{U}),\overline{\mathsf{E}}}$ is boundedly retractive by our conclusion 7 above. This is easily seen to imply that the direct limit topology is sequentially retractive, i.e., that convergent sequences are contained in, and convergent in, a component of the direct limit [6]. This implies that there exists $j\in\mathbb{Z}_{>0}$ such that the sequence $(\overline{\xi}_k)_{k\in\mathbb{Z}_{>0}}$ converges in $\mathrm{L}^1(\mathbb{T}';\Gamma^{\mathrm{hol},\mathbb{R}}(\overline{\mathsf{E}}|\overline{\mathcal{U}}_j))$ and so converges to a limit $\overline{\eta}$ satisfying $[\overline{\eta}]_{\mathrm{cl}(\mathcal{U}_j)} = [\overline{\xi}]_{\mathrm{cl}(\mathcal{U}_j)}$. Thus $\overline{\xi}$ can be holomorphically extended to $\overline{\mathcal{U}}_j$. This completes this part of the proof.

(ii) Let $K\subseteq\mathsf{M}$ be compact and let $\boldsymbol{a}\in c_0(\mathbb{Z}_{\geq 0};\mathbb{R}_{>0})$. Let $(\overline{\mathcal{U}}_j)_{j\in\mathbb{Z}_{>0}}$ be a sequence of neighbourhoods of K in $\overline{\mathsf{M}}$ such that $\mathrm{cl}(\overline{\mathcal{U}}_{j+1})\subseteq\overline{\mathcal{U}}_j$ and $K=\cap_{j\in\mathbb{Z}_{>0}}\overline{\mathcal{U}}_j$. By hypothesis, for $x\in K$, there is a relatively compact neighbourhood $\overline{\mathcal{U}}_x\subseteq\overline{\mathsf{M}}$ of x in $\overline{\mathsf{M}}$ such that there is an extension $\overline{\xi}_x\in\mathrm{LI}\Gamma^{\mathrm{hol},\mathbb{R}}(\mathbb{T};\overline{\mathsf{E}}|\overline{\mathcal{U}}_x)$ of $\xi|(\mathbb{T}\times(\overline{\mathcal{U}}_x\cap\mathsf{M}))$. Let $x_1,\dots,x_k\in K$ be such that $K\subseteq\cup_{j=1}^k\overline{\mathcal{U}}_{x_j}$ and let $l\in\mathbb{Z}_{>0}$ be sufficiently large that $\overline{\mathcal{U}}_l\subseteq\cup_{j=1}^k\overline{\mathcal{U}}_{x_j}$, so ξ admits an holomorphic extension $\overline{\xi}\in\mathrm{LI}\Gamma^{\mathrm{hol},\mathbb{R}}(\mathbb{T};\overline{\mathsf{E}\mathcal{U}_l})$.

Now we show that the above constructions imply that $\xi \in \mathrm{L I}\Gamma^{\omega}(\mathbb{T};\mathsf{TM})$. Let $\overline{g} \in \mathrm{L}^1_{\mathrm{loc}}(\mathbb{T};\mathbb{R}_{\geq 0})$ be such that

$$\|\overline{\xi}(t,z)\|_{\overline{\mathbb{G}}} \leq \overline{g}(t), \qquad (t,z) \in \mathbb{T}\times\overline{\mathcal{U}}_l.$$

By Proposition 4.2, there exist $C, r \in \mathbb{R}_{>0}$ such that

$$\|j_m\xi(t,x)\| \leq Cr^{-m}\overline{g}(t)$$

for all $m \in \mathbb{Z}_{\geq 0}$, $t \in \mathbb{T}$, and $x \in K$. Now let $N \in \mathbb{Z}_{\geq 0}$ be such that $a_{N+1} < r$ and let $g \in \mathrm{L}^1_{\mathrm{loc}}(\mathbb{T};\mathbb{R}_{\geq 0})$ be such that

$$Ca_0a_1\cdots a_m r^{-m}\overline{g}(t) \leq g(t)$$

for $m \in \{0,1,\dots,N\}$. Now, if $m \in \{0,1,\dots,N\}$, we have

$$a_0a_1\cdots a_m\|j_m\xi(t,x)\|_{\overline{\mathbb{G}}_m} \leq a_0a_1\cdots a_m Cr^{-m}\overline{g}(t) \leq g(t)$$

for $(t,x) \in \mathbb{T}\times K$. If $m > N$ we also have

$$\begin{aligned} a_0a_1\cdots a_m\|j_m\xi(t,x)\|_{\overline{\mathbb{G}}_m} &\leq a_0a_1\cdots a_N r^{-N}r^m\|j_m\xi(t,x)\|_{\overline{\mathbb{G}}_m} \\ &\leq a_0a_1\cdots a_N r^{-N}C\overline{g}(t) \leq g(t), \end{aligned}$$

for $(t,x) \in \mathbb{T}\times K$, as desired. □

Finally, let us show that, according to our definitions, real analytic time-varying vector fields possess flows depending in a real analytic way on initial condition.

Theorem 6.26 (Flows of Vector Fields from $\mathrm{L I}\Gamma^{\omega}(\mathbb{T};\mathsf{TM})$). *Let* M *be a real analytic manifold, let* $\mathbb{T}$ *be an interval, and let* $X \in \mathrm{L I}\Gamma^{\omega}(\mathbb{T};\mathsf{TM})$. *Then there exist a subset* $D_X \subseteq \mathbb{T}\times\mathbb{T}\times\mathsf{M}$ *and a map* $\Phi^X : D_X \to \mathsf{M}$ *with the following properties for each* $(t_0,x_0) \in \mathbb{T}\times\mathsf{M}$*:*

(i) the set

$$\mathbb{T}_X(t_0,x_0) = \{t \in \mathbb{T} \mid (t,t_0,x_0) \in D_X\}$$

is an interval;

(ii) there exists a locally absolutely continuous curve $t \mapsto \xi(t)$ *satisfying*

$$\xi'(t) = X(t,\xi(t)), \quad \xi(t_0) = x_0,$$

for almost all $t \in |t_0,t_1|$ *if and only if* $t_1 \in \mathbb{T}_X(t_0,x_0)$*;*

(iii) $\frac{\mathrm{d}}{\mathrm{d}t}\Phi^X(t,t_0,x_0) = X(t,\Phi^X(t,t_0,x_0))$ *for almost all* $t \in \mathbb{T}_X(t_0,x_0)$*;*

(iv) for each $t \in \mathbb{T}$ *for which* $(t,t_0,x_0) \in D_X$*, there exists a neighbourhood* $\mathcal{U}$ *of* x_0 *such that the mapping* $x \mapsto \Phi^X(t,t_0,x)$ *is defined and of class* C^{ω} *on* $\mathcal{U}$*.*

Proof: The theorem follows from Theorems 6.18 and 6.25, noting that the flow of an holomorphic extension will leave invariant the real analytic manifold. □

6.5 Mixing Regularity Hypotheses

It is possible to mix regularity conditions for vector fields. By this we mean that one can consider vector fields whose dependence on state is more regular than their joint state/time dependence. This can be done by considering $m \in \mathbb{Z}_{\geq 0}$, $m' \in \{0, \text{lip}\}$, $r \in \mathbb{Z}_{\geq 0} \cup \{\infty, \omega\}$, and $r' \in \{0, \text{lip}\}$ satisfying $m + m' < r + r'$, and considering vector fields in

$$\mathrm{CF}\Gamma^{r+r'}(\mathbb{T}; \mathsf{TM}) \cap \mathrm{LI}\Gamma^{m+m'}(\mathbb{T}; \mathsf{TM}) \quad \text{or} \quad \mathrm{CF}\Gamma^{r+r'}(\mathbb{T}; \mathsf{TM}) \cap \mathrm{LB}\Gamma^{m+m'}(\mathbb{T}; \mathsf{TM}),$$

using the obvious convention that $\infty + \text{lip} = \infty$ and $\omega + \text{lip} = \omega$. This does come across as quite unnatural in our framework, and perhaps it is right that it should. Moreover, because the $\mathrm{CO}^{m+m'}$-topology for $\Gamma^{r+r'}(\mathsf{TM})$ will be complete if and only if $m + m' = r + r'$, some of the results above will not translate to this mixed class of time-varying vector fields: particularly, the results on Bochner integrability require completeness. Nonetheless, this mixing of regularity assumptions is quite common in the literature. Indeed, this has *always* been done in the real analytic case, since the notions of "locally integrally C^{ω}-bounded" and "locally essentially C^{ω}-bounded" given in Definition 6.20 are being given for the first time in this monograph.

References

1. Abraham, R., Marsden, J.E., Ratiu, T.S.: Manifolds, Tensor Analysis, and Applications, 2 edn. No. 75 in Applied Mathematical Sciences. Springer-Verlag (1988)
2. Agrachev, A.A., Gamkrelidze, R.V.: The exponential representation of flows and the chronological calculus. Mathematics of the USSR-Sbornik **107**(4), 467–532 (1978)
3. Agrachev, A.A., Sachkov, Y.: Control Theory from the Geometric Viewpoint, *Encyclopedia of Mathematical Sciences*, vol. 87. Springer-Verlag, New York/Heidelberg/Berlin (2004)
4. Beckmann, R., Deitmar, A.: Strong vector valued integrals (2011). URL http://arxiv.org/abs/1102.1246v1. ArXiv:1102.1246v1 [math.FA]
5. Bogachev, V.I.: Measure Theory, vol. 2. Springer-Verlag, New York/Heidelberg/Berlin (2007)
6. Fernández, C.: Regularity conditions on (*LF*)-spaces. Archiv der Mathematik. Archives of Mathematics. Archives Mathématiques **54**, 380–383 (1990)
7. Fritzsche, K., Grauert, H.: From Holomorphic Functions to Complex Manifolds. No. 213 in Graduate Texts in Mathematics. Springer-Verlag, New York/Heidelberg/Berlin (2002)
8. Gunning, R.C.: Introduction to Holomorphic Functions of Several Variables. Volume I: Function Theory. Wadsworth & Brooks/Cole Mathematics Series. Wadsworth & Brooks/Cole, Belmont, CA (1990)
9. Hewitt, E., Stromberg, K.: Real and Abstract Analysis. No. 25 in Graduate Texts in Mathematics. Springer-Verlag, New York/Heidelberg/Berlin (1975)
10. Jost, J.: Postmodern Analysis, 3 edn. Universitext. Springer-Verlag, New York/Heidelberg/-Berlin (2005)
11. Mangino, E.M.: (LF)-spaces and tensor products. Mathematische Nachrichten **185**, 149–162 (1997)
12. Retakh, V.S.: The subspaces of a countable inductive limit. Soviet Mathematics. Doklady. A translation of the mathematics section of Doklady Akademii Nauk SSSR **11**, 1384–1386 (1970)

13. Saunders, D.J.: The Geometry of Jet Bundles. No. 142 in London Mathematical Society Lecture Note Series. Cambridge University Press, New York/Port Chester/Melbourne/Sydney (1989)
14. Schaefer, H.H., Wolff, M.P.: Topological Vector Spaces, 2 edn. No. 3 in Graduate Texts in Mathematics. Springer-Verlag, New York/Heidelberg/Berlin (1999)
15. Schuricht, F., von der Mosel, H.: Ordinary differential equations with measurable right-hand side and parameter dependence. Tech. Rep. Preprint 676, Universität Bonn, SFB 256 (2000)
16. Sontag, E.D.: Mathematical Control Theory: Deterministic Finite Dimensional Systems, 2 edn. No. 6 in Texts in Applied Mathematics. Springer-Verlag, New York/Heidelberg/Berlin (1998)
17. Sussmann, H.J.: An introduction to the coordinate-free maximum principle. In: B. Jakubczyk, W. Respondek (eds.) Geometry of Feedback and Optimal Control, pp. 463–557. Dekker Marcel Dekker, New York (1997)
18. Thomas, G.E.F.: Integration of functions with values in locally convex Suslin spaces. Transactions of the American Mathematical Society **212**, 61–81 (1975)
19. Wengenroth, J.: Retractive (LF)-spaces. Ph.D. thesis, Universität Trier, Trier, Germany (1995)
20. Whitney, H.: Differentiable manifolds. Annals of Mathematics. Second Series **37**(3), 645–680 (1936)